KEEPING HORSES OUTDOORS

Smart, Safe, Affordable Ways Your Horse Can Live as Naturally as Possible All Seasons of the Year

Iveta Jebáčková-Lažanská

Translated by Michaela Klofac | Edited by Sarah O'Neill

Trafalgar Square
North Pomfret, Vermont

First published in the United States of America in 2024 by
Trafalgar Square Books
North Pomfret, Vermont

Originally published in Czech as *Venkovní Ustájení Koní*.

Disclaimer of Liability
The author and publisher shall have neither liability nor responsibility to any person or entity with respect to any loss or damage caused or alleged to be caused directly or indirectly by the information contained in this book. While the book is as accurate as the author can make it, there may be errors, omissions, and inaccuracies.

Trafalgar Square Books encourages the use of approved safety helmets in all equestrian sports and activities.

Trafalgar Square Books certifies that the content in this book was generated by a human expert on the subject, and the content was edited, fact-checked, and proofread by human publishing specialists with a lifetime of equestrian knowledge. TSB does not publish books generated by artificial intelligence (AI).

ISBN: 978-1-64601-108-7
Library of Congress Control Number: 2024941067

Czech editor: Šárka Dolanská
Cover design: RM Didier
Interior graphic design: Tomáš Marek, Daniel Burda
Translation into English: Michaela Klofac and Sarah O'Neill
Index by Andrea M. Jones (www.jonesliteraryservices.com)

Printed in China
10 9 8 7 6 5 4 3 2 1

Contents

Preface

Since I began creating a system for keeping horses outside, I have seen great enthusiasm for the need to return horses to a more natural state where continuous movement is possible and where they enjoy the comfort of being part of a herd. From my childhood, I can clearly remember horses being confined for 23 hours or more due to inclement weather or a lack of turnout space. Today, I want my horses to have the best possible access to outdoor space.

What bonds those of us who keep our horses outside is our shared efforts to convince others who may subscribe to an outmoded school of thought of the immense benefits of having horses allowed a life "under the open sky."

I support all horse people who make an effort to provide their horses with free-choice stalls and even a couple hundred square feet of outdoor space—some is better than none! I applaud every open door in a stable. Unfortunately, the response from many owners of sport horses has not been encouraging—while those who own "backyard horses" accept that they can be kept outside, it seems harder for those with horses they compete. An equine athlete left outdoors in freezing temperatures rather than under a roof in a warm stall? Absolutely not!

The belief that keeping horses outside was hugely inconvenient turned many horse owners away from the practice. Horsemen of the past often thought the cold and rain would surely sicken horses, and that was enough for most to never venture past the four walls of a stall. Others alleged the horse was unable to rest sufficiently after competition or travel when left outdoors. Still others were disgusted by mud or long, cold winters—they clearly saw difficulty in caring for horses outdoors. And so, the less determined left the fight to keep the horse naturally in favor of the tidy indoor stall. Others persisted. And those who have slowly but surely found ways to create a healthy outdoor situation for their horses.

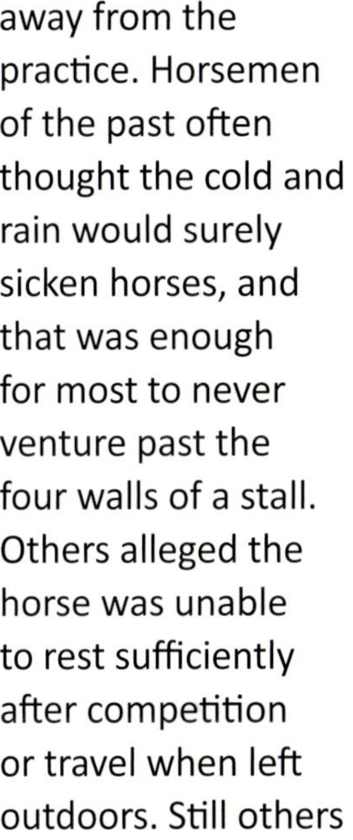

The beginnings of this movement have been tough for some! Especially those with only small amounts of land, who really have to work hard to make it work. No one likes to see horses trapped ankle-deep in mud from autumn until spring, with no dry area to escape to.

But now, for those who want to, it's possible to find ways to make outdoor horsekeeping work. After years of simple "blind enthusiasm," a system has slowly emerged that can work for any property, whether small or large. It has been proven to work well by horse owners around the world.

The following pages reflect recommendations from this system, as I share my own experiences, and also those of my friends with farms and horse properties, large and small. View this book as providing examples of the paths some have taken, not as a command to "drive this way, no turning!"

This book, *Keeping Horses Outdoors*, is for all eager, hard-working, curious horse owners who want their horses to live as naturally as possible in outdoor spaces and are desperately looking for advice and guidance. I believe that readers will appreciate learning from the skilled horsemen who generously shared not only photos of their carefully built properties, but also their own experiences. Many, many thanks to all these contributors, without whom this book could not have been made. I thank them wholeheartedly.

Dear readers, I hope you find this book useful.
Iveta Jebáčková-Lažanská

About the Author

Iveta Jebáčková-Lažanská has been active in the horse world for almost four decades. As a child, she started in the equestrian discipline, where she also got to know the operational side of keeping horses—at a time when horses still spent most of the day locked in stalls (at best; at worst, they stayed tied to a stake). Even then, she was considering the possibilities of building an environment in which horses could function on their own, without any need for constant attention or supervision by people, without whom the standard design of a stable means they are condemned to stand around idly in confined spaces.

Later in her life, she and her husband began to slowly build facilities of their own, and later stabled animals on their farm, keeping the gates wide open to encourage free movement by the horses. She regularly writes articles about her experiences with outdoor stables for Czech horse websites and magazines, and also writes romances with horses for young people, especially girls (she herself has three daughters).

With this book, she wants to help everyone who is new to housing horses outdoors and is looking for a sophisticated, functional system in which horses, nature, and people thrive.

The author's farm (both photos).

chapter I.

KEEPING HORSES IN A MUD-FREE ENVIRONMENT YEAR-ROUND

More and more, large equestrian clubs and stud farms are being replaced by smaller barns. As smaller stables have become more commonplace, the popularity of increased turnout has risen accordingly—not just locally, but around the world. Popularity aside, it's expensive and time-consuming to create a comfortable environment for horses who live turned out. The biggest problem? Mud.

If you have horses, mud is unavoidable. But there's a huge difference between a horse trapped in a small muddy space and a horse who only has to pass through a small wet section in a large and otherwise dry paddock. Dry conditions don't only have to be for huge, multi-million dollar facilities. It can be done for the backyard outdoor horse. Ideal stabling should incorporate the best of both worlds—a comfortable stall and freedom to roam and graze.

The primary goals when creating an outdoor space are:

- Some consideration for human comfort, which might not be the priority at the start but can gradually be incorporated over time as work on the space progresses. No one wants to kneel in the mud, or make a vet trudge through a swamp. "Comfort" for humans means at least some access to dry places.
- Comfort for the horse, which should initially, at a minimum, ensure access to some mud-free areas, and to enough space to ensure that less dominant members of a herd don't have to fight for access to food or shelter.

Pay attention to the number of horses in an area. Respect the need for peace in the herd.

Horses walking freely between paddock and pasture.

Ways to make life easier for you and your horses:

- Easy access to electricity and clean water.
- Sheltered areas with pavers or stone dust for when it rains.
- Sufficient space for unlimited time outside (of course, "sufficient" takes on a different meaning in the suburbs outside a city than in the countryside).
- A quarantine area for sick or new horses: at least two quality box stalls (in case a companion horse is needed).

Keep in mind how many horses your land can accommodate when conditions aren't ideal; if it's raining and they can't use all of the space available, will your horses be cramped, or can they be comfortable?

The best thing about keeping horses outdoors is giving them *freedom of choice*. As their caretakers, we support them by appropriately dividing the herd to assure that they are calm and happy, and by providing them with enough options so they can move freely and problem-solve for themselves. It's our responsibility to provide our horses with sufficient options so they may seek comfort as needed by providing relief from heat, rain, cold, and wind (for example, run-in shelters, natural terrain which can serve as a windbreak, tree cover, or a combination).

It's worth mentioning that giving our horses these options has another important benefit: reducing the need for constant care. **With something as simple as a well-placed run-in, we can eliminate the need to bring horses in and out of their stalls each day and with each weather change.**

All of this, of course, is dependent on the amount of space a farm manager can offer: some may have the freedom to provide their animals with twenty acres, others with only one or two. Either way, the common denominator should always be that our horses are given as much freedom and autonomy as space will allow. We can all agree that a closed stall offers no freedom of choice whatsoever.

One of the benefits of outdoor housing is the peace of mind it can lend to the horse's human! A horse who is able to interact with a pal or two and move around throughout his day will be happier, and his owner will have so much less to worry about. Knowing "the horse is fine without me" is a wonderful feeling. An owner who's had a busy day and hasn't been able to work with a horse they keep outside will rest much easier than someone who doesn't have the time for a horse kept in a confined space, staring at his buddies in vain through the bars of his stall door.

Nature did not intend for horses to be kept in small spaces: their musculoskeletal, respiratory, and digestive systems evolved around a continuous, gradual intake of forage as they travel over a wide area. Horses are designed to graze, so they are happiest when they can move constantly and eat with their heads lowered to the ground. Horses also feel the most secure when they are part of a group, not placed in individual stalls or paddocks. Many owners these days are clearly interested in trying to simulate a more natural environment for their horses, with well-planned runs and pastures set up to allow horses to behave as Mother Nature intended whenever they are not being handled—just look at online discussion groups.

Outdoor horse keeping allows owners who can't ride regularly to keep their horses fit and happy.

Horses at Home

Because many of our horses are for pleasure riding and companionship first and foremost, the number of people who want to keep their horses at home is growing at a fast pace. More and more people are choosing to keep their horses in their backyards, where they can make decisions about their needs and care, rather than put up with the necessary compromises that come with boarding at a commercial stable.

If there is only room for a small turnout connected to a shelter, it is absolutely necessary to be careful about choice of footing and drainage.

Many find it calming and reassuring to care for their own horses. However, the demands of full-time jobs and the often time-consuming commute from work to barn can make this difficult, which has also contributed to the number of equestrians moving their horses home.

But how can you ensure that your backyard horses are truly happy? That they aren't standing in wet shavings and mud? That they are eating enough, but not overeating? That they are taking advantage of all the space available to them, and not just hovering around a feeder? That they aren't bothered too much by insects? That they aren't too hot or cold? If you're asking these questions, you're also looking for solutions. This book aims to help!

Outdoor or Pasture?

I'd like to define the terms I'll use regularly in this book. Specifically, the words "outdoor" and "pasture" as I'm going to use them here aren't about the amount of hay or grass a horse is fed, but rather how and when different spaces are used.

Outdoor housing/stabling—I use this phrase to discuss a set-up where horses are primarily kept in outdoor cultivated areas with basic enclosures or shelter, supplemented by pastures, forest, or a combination.

Horses in this situation are allowed on pasture only during dry or frozen conditions; otherwise, they are kept in basic paddocks.

Pasture stabling—A system for keeping horses year-round in large pastures, which are used indefinitely for grazing and rotating animals regardless of the temperature. In this system, basic paddocks are mostly not used.

Frequently Asked Question:
Should we bring our horses home and work on the spaces they're going to use a bit at a time? Or would it be better to leave them boarded for a year while we set things up, so they will arrive at a finished stable?

If you can manage the financial burden of paying board as well as paying for the cost of construction, the preparation of turnout spaces and shelters is much easier if you don't have to work around your horses. If you do have the choice, try to keep your horses at their current location until you have prepared at least a basic shelter with a fenced-in area surrounding it.

The rest can be finished gradually, after moving your horses in, but if you don't have at least a basic

Demand for outdoor housing is growing each year.

A working farm in Germany. These horses have halters with computer chips in them, allowing them to gain access to feeders that distribute a precisely set amount of feed.

run-in for the animals, that's not an ideal start. And if construction can be done without any need to worry about the horses during the process, it will help speed things along.

Choosing land:

Everyone who buys, inherits, or builds a house with the intention of keeping horses should consider what purpose the horses' housing will serve, and act accordingly:

- numbers and kinds of animals
  or
- type of land

If you are shopping for land and have the chance to make a choice—choose well! It's sad to see small, cramped areas holding a large number of animals;

In this overwhelmingly technical age we live in, the sight of happy horses demonstrably "recharges our batteries."

often the owners started with the belief that "the space behind the house is enough, we'll only have two horses," but then ended up with ten...

Know **how many and what kind of horses** you intend to keep ahead of time. For example, having a couple retired horses who may find it difficult to move or have additional needs, and will require extra care, means finding housing that allows for easy access to these "old-timers," in case you have to give them regular medication or other assistance.

Always consider the future as well—a small adjacent property may work well right now for those retirees, but will they someday be replaced by younger horses who will need more space?

If, on the other hand, there's no choice to be had, the space will dictate what can be done with it. For example, say you inherit a small house with a little garden and a single back field; you can't insist on keeping a herd of broodmares and a stallion to boot!

Real-life advice: don't count on leased land! If you have an agreement to use someone else's land, be sure you have enough space of your own to keep your horses yourself, should you suddenly lose the use of that property.

Less space is needed for actually riding horses, whether you are preparing for lower-level competitions or regular trail rides in surrounding areas.

A 20 x 40-meter paddock should be adequate, if owners ride at least every other day. Horses will experience the amount of movement they need just by going from strategically placed hay to water, to covered space.

Having horses at home and allowing them all-day turnout is not just a trend, but is gaining in popularity everywhere.

If a sandy area is sloped, think about where the run-off will end up after it rains. You'll save yourself a lot of work cleaning up if paved surfaces sit higher than sandy ones.

On the other hand, broodmares with foals not only require a large amount of space, but also need regular access to good pasture.

Frequently Asked Question:

I worry about vandalism in the fields behind our house. Not far from our pasture is an area that kids use in the evening. On many mornings, I've found beer cans, trash, and once even a fence burnt by a lighter. I'm afraid to leave the horses unattended all night long with just an electric fence. Is it necessary to keep them indoors at night?

It is not necessary!

If you have to keep the safety and security of your horses at night in mind, you should have access to safe land where the animals are within easy reach of a stable or owner's house, and can be supervised at all times. Use proximity to manage their safety, not confinement! If you want to keep horses outside all year because you believe it will keep them their healthiest and happiest, then closing them up night after night in a 10 x 12 stall runs completely contrary to everything you're working to achieve.

Studies show that horses who are able to move consistently throughout the day and night are more focused, learn better, and are more willing to work with their handlers. Also, the health benefits are significant.

chapter II.

CHARACTERISTICS OF BASIC PADDOCKS

A basic paddock is a simple enclosure: a relatively small, safely fenced-in area that provides enough room to avoid skirmishes among the animals and is usable year-round. It should be near stabling or your house, and have enough variety in the footing to keep mud to a minimum. The horses should have easy access to hay, clean water, space to comfortably lie down, and covered or leeward areas to shelter them from bad weather.

Why Set Up a Basic Paddock?

Because sometimes it's necessary to put horses in a smaller space, whether it's only a few times per year or every night. Whether you need a few days after deworming to collect manure, or after a heavy rain to protect soaked turf from damage thanks to too many hooves, it can be fantastic to have a well-planned-out, functional space at your disposal. Sure, you also have stalls, but why should your horses spend an entire week in a stall? There's no reason they should have to, if you've built a paddock for them!

Building a good paddock doesn't mean following precise rules—as long as the necessities are accounted for, each owner can decide how to arrange the space and what else to include.

What this area needs to contain:

- electricity
- a dry area to feed hay and grain
- windbreaks
- shelter from rain
- space to lie down
- salt and mineral blocks

In addition to the necessities above, it should have:

- areas for horses to scratch and roll
- varied footing surfaces

I'll discuss each of these elements in more detail later on.

The advantages of this type of area over a stall are:

- Providing much greater freedom of movement.
- Allowing unimpeded social interaction between horses.
- Allowing horses the freedom to choose between shelter or outdoor spaces.
- Giving horses the ability to move over varied surfaces, day and night.
- Giving horses far greater mental stimulation.
- Providing much more space, in even the smallest run, than in even the largest stall.

The advantages of a small paddock as a supplement to pasture:

- This setup is less demanding for us humans, especially during "mud season."
- Pastures can be rested to avoid expensive damage to turf.
- It's possible to provide better security (if needed) for horses at night.
- Horses can be removed from contaminated grass, especially if neighboring land is agricultural and gets sprayed with pesticides at specific times.
- Horses can be quarantined or kept close for supervision (for example, after deworming).

If you build truly spacious shelters for your horses, there's no doubt they'll be happy to use them.

When there is bad weather that lasts for several days, if horses have access to a paddock, we can rest assured that they can be out of their stalls and moving without any risk they'll trample their pastures into mud. Being able to move freely day and night, as well as being able to choose where and with whom he spends his time, is extremely beneficial to a horse's well-being. Being able to choose is important for horses (as for any living creature) in all areas of life. According to studies, this type of self-expression stimulates brain activity and evokes relaxation. If a horse is free to choose where he will go and how long he will stay there (whether he decides to be under a roof, in open space, alone around the corner, or in the center of activity), his psychological resilience—which is to say, his ability to react calmly to stimuli—rises significantly. It's up to each barn owner to decide whether to offer a basic paddock for the group without stalls at night (thus reaping all of the aforementioned benefits), or a stone dust paddock with an adjoining shed row where horses can be enclosed as a group or individually. Whether the shed row in the latter option is used occasionally or as part of a daily routine (for example, to feed grain), having horses accustomed to a bit of privacy can be great for horse and human!

If pasture will be used for both grazing as well as hay for the winter, the required size is as follows:

2½ acres = 6 retired horses, three yearlings or 1 adult horse
(Doležal, Navrátil; 1994)

Nowadays, however, rainfall amounts in some areas are decreasing, and many farmers without basic paddocks are finding that even more land is needed to maintain the quality of their pasture. Four or five acres is an ideal amount of space for a single adult horse, but in extreme drought conditions, quite a bit more is needed. This holds true especially if the horses' diet will depend heavily on grazing. In this case, there must be an organic soil profile with a healthy mixture of microorganisms.

If ample hay is provided and pasture space is used primarily to promote movement, one acre is enough grazing space for each adult horse.

These are, of course, ideal circumstances. In reality, we can find pleasant, functional outdoor stabling even on much smaller spaces, where two or three horses are thriving on only an acre. This is possible with thoughtfully designed space and a close-knit herd. In a larger facility with greater turnover, the herd is ever-changing, which creates nervousness and emotional pressure. In a system like this, a much greater amount of space, allowing horses to be separated, is the only way to prevent total chaos.

Some handlers prefer to use stalls with open access to runs rather than basic paddocks.

On the other hand, a close-knit herd, provided they have adequate exercise, will be a calm group that can live in peace even if circumstances require they be put in a smaller space temporarily.

Mud Is the Enemy

Repeat this mantra: Mud is the enemy. Lots of folks in the industry will tell you that mud is a normal part of having horses, and that it's unavoidable. On the one hand, they're right. But on the other hand, they're wrong... Why is that?

Ever since animals were first domesticated and confined in man-made structures rather than grazing freely on vast expanses of land, people have had to solve the problems caused by mud and manure. Anywhere animals are confined outdoors, the inevitable mixture of manure, dirt, and water creates an ideal environment for bacteria.

These soggy spaces couldn't be used for more than a few hours, and certainly not long-term. So, in order to protect their animals not only from predators and thieves, but also from mud, people began to enclose horses in stables. As populations increased, so did the number of horses under saddle and in harness. People began to notice the condition of streets and the yards around houses. At first, stables were not an aesthetic choice, but a practical one; horses walking on dry ground were less likely to damage it. In areas with lots of rain, the easiest way to prevent such large animals from doing a great deal of damage to the land was to confine them to solid indoor footing. The easiest way to build and maintain a dry space was to make it as small as possible. And so stables were created...

So, if people try to tell you that in your smaller spaces and paddocks, mud simply needs to be endured, don't believe them. Stables were invented to avoid mud, and there are ways to do the same in a well-planned paddock.

Here is an example of two interconnected paddocks. One is used for feeding and the other for shelter, and the horses can pass freely between them.

Historically, the effort to avoid mud in horsekeeping was not for the sake of cleanliness, or because of the threat of infection, but rather to make work easier. In knee-high mud, work around cattle and horses only became more difficult.

In this space, horses can choose between different types of feeding stations.

Horses can still be horses without mud. Don't get caught up wishing you had a huge, naturally dry paddock like your neighbor does. Instead, focus on your own space and what you can do with it. You can probably accomplish more than you think.

Picture a field divided into squares the size of a horse's hoof. If the field is large, the probability that a horse will step in the same square more than once over the course of a day is pretty small. However, if the field isn't as big, a horse will probably place his hoof in most or even all of the squares—and not just once or twice, but a dozen times, or a hundred. Now, imagine this is happening on a rainy day. Grass becomes mud.

In many areas, stables with less than an acre per horse are common. It's easy to imagine a single healthy horse turning that acre to mud in short order on a wet autumn day. This is why it can be so hard to shake the "conventional wisdom" that where horses go, mud follows.

The owners of large properties will continue to enjoy mud only at the gates to their fields. However, owners of smaller spaces with limited pasture face a difficult situation; if they don't want to confine their horses to their stalls in order to rest the fields, they may think they have to sacrifice some of their space to mud. This can turn into a huge problem—a water-soaked pit of manure can form where horses can injure themselves from becoming quite literally stuck in the mud.

If we want to keep our pasture in the best possible condition, we should keep our horses off it during wet weather, ideally until it's completely dry. So, in order to save our fields and still have happy horses, a good basic paddock is a necessity.

The Basic Paddock: The Ideal Size

The size of this space depends on two main things: the number of horses, and their ages. We need to keep in mind that a basic paddock needs to be big enough to accommodate the horses any time they need to be taken off pasture. The size you need will depend greatly on how you expect to use it. There's an obvious difference between a paddock you are only going to use occasionally, and a paddock your horses will be in every night. I can't offer you truly precise measurements here, just estimates. Anything larger than a stall is, of course, an improvement—but safety must be taken into account. Horses shouldn't run up against a fence line after two canter strides...

Here we come to our first definition: **a basic paddock is not a run.** Runs, despite their name, don't allow horses to run. A horse confined to a stall with a run knows he's in a space where he can't do more than walk or trot some, simply because the size of the area doesn't allow it. A basic paddock, on the other hand, will allow smaller horses to run without any difficulty.

For two calm, older horses, a basic paddock of 20 x 20 meters should be sufficient for short rainy periods off pasture, provided the horses are exercised and allowed back in their fields as soon as the ground has dried. However, for younger animals, more physically active breeds, or larger groups, any enclosure should be large enough to allow them to move freely.

A small, crowded space breeds unnecessary aggression. A larger area means happier, well-adjusted horses.

When the space you create for your horses is large enough, it can significantly improve the herd dynamic. But it won't be cheap!

Mud:

- Is going to show up here and there, but shouldn't be everywhere!
- Is found in nature, especially near watering holes. For horses in confinement, however, mud can become dangerous to their health when it mixes with urine and manure. Our horses also have nowhere to escape to, and they shouldn't have to stand in mud all day.
- Is a breeding ground for insects and bacteria.
- Affects soil's ability to absorb water, and prevents the growth of plants which would otherwise handle changes in moisture (wet to dry and back again) better than grass.
- Increases the risk of accidents and illness.
- Can form an uneven and painful surface when it freezes, creating the risk of abcesses, strained tendons and ligaments, and even acute trauma such as fractures. Colic and digestive upset is another possibility: some horses will be unwilling to walk across frozen muddy areas in order to drink or eat.

There's no question that young horses will run, at least occasionally, in their paddock. In order to avoid frayed nerves over the constant worry that our horses might not be able to stop at the fence line, or that they might run through electric tape on a windy day, we need to not only think about the size, but the entire set-up of this space, always making sure the safety of the horse is the number one priority.

Ideal dimensions for this space will allow all the horses who are going to use it to exist in harmony; they should be able to share it without feeling cramped or unable to escape other members of the herd. Each horse should be able to find a moment or two of privacy—around the corner, behind a shelter or bush, and so on.

A basic paddock, with the footing reinforced with heavy mats.

Stall with a run-out. A freely accessible area adjoining the stall allows for greater comfort for the horse, but is too small to allow him to run.

Whatever you build, it will be specific to your needs and those of your animals. The specifics will depend not only on the age and temperament of your individual horses and how well they interact with each other, but also on what they are used to. For example—50 x 50 meters will look very different to an animal accustomed to wide open fields than it will to a young horse who's lived in a stall since birth.

It's up to you, the owner, to choose a sensible compromise that will allow your horses as much space as possible, but is also functional and financially attainable for you. The larger this enclosure can be, the more happily members of the herd can coexist, but if you don't have the resources to invest in correct preparation for a larger surface, it's better to focus on a smaller space that will allow you to afford to build correctly and achieve the right quality.

There's no need to provide your horses with so much room in their paddock that they can pick up speed and race. In fact, this could result in injury. But anyone who wants to call themselves a true horseperson should be able to provide a space where horses can trot or canter a 20-meter circle.

Practical Observations

If you build a paddock that's truly functional, you'll know it! The horses will love it, and will express that love by choosing to come into their paddock from the fields all the time. Horses should feel safe in this space, and should want to choose this area to rest in.

My horses view their paddock as a safe haven, and associate it with all kinds of pleasant and interesting activities. This is where they can always find hay and water, a space to cool off on hot days, or a person arriving with a curry comb or a treat. If something startles my horses when they're out in their pasture, the herd comes rushing into the paddock. Even a casual observer will notice how quickly the horses relax in the oasis of this enclosure, where they find peace and security.

It's important to note that this space should provide options for shelter as well. My "grandfather" gelding has shown me repeatedly that he views a large shed I've provided as a safe space: each time a new horse is introduced, he drives all the mares with foals into the shed while he assesses the "threat" this new intruder poses. I've seen this happen several times, and it's always heartwarming to see how happy and safe the horses feel in their paddock, and specifically the shelter I created for them inside it.

chapter III.

SHELTER

The most important element of our basic paddock is a spacious shelter—a dry, covered place in which our horses can find relief from rain, wind, and sun. Since this shelter has to protect from wind as well as sun and rain, it must have several solid walls in addition to a roof. It's smart to keep in mind the direction of any seasonal winds in your area when you're choosing a location, but in some places, wind comes from every direction. If this is the case where you are and the front of the shelter you create is entirely open, horses won't always be able to find a protected place to rest.

So how can you create shelter from wind that blows from any direction? Let's build a run-in shed that horses can go in or around, and, at least for certain months, we can add an upwind wall, placed in the middle of the shelter and extending one-third or half of the way across the length of the shed.

Some people also opt for windbreaks at the entrance to the shelter, but in my experience these get mixed reviews. They work fine for some, but in truly high winds, they don't always do very much. Nothing is so big and solid it can't be blown away.

If your run-in will only be used seasonally, for example, in a pasture with tree cover, the overall length and width don't matter as much, so long as it's safe. Since these sheds won't be used as often, a lower height is fine as well.

However, the shelter you create in your paddock is another matter entirely. This shelter should function as a space where any activity we would perform in the paddock can also be done. The shelter should be large enough to allow for handling and feeding during inclement weather. Horses should be able to retreat to this space from adjoining pastures whenever they need to. To make this central space truly functional, the main feature—the sheltered space—must be genuinely serviceable. The main shelter should be as large and as high as possible. In the summer months, any additional height is a plus; a high roof with a gap or vents will allow hot air to escape much better than a lower ceiling.

Instead of providing shelters, some people would rather put rain sheets on their horses. I prefer to invest in a structure that works as both sunblock and umbrella, rather than constantly tracking the weather and racing to beat it.

Strip curtain doors—popular with some, disliked by others...

Strategic Placement of Shelter

Dry Feet All the Way

If you have to push a wheelbarrow through deep mud several times a day during mud season, not only will your body pay the price, you'll also learn to hate this time of year (which, in some areas, can be almost year-round). It's pointless to work so hard; you're only wearing yourself out and discouraging anyone else from pitching in more than once. It's far better to think about future manual labor during the building process. Our predecessors, being accustomed to the hard work of maintaining livestock, had barns and farms organized around the daily work they knew they would need to do, and made as few treks through the mud and elements as possible.

The Manure Pile: Accessibility

It would be lovely to be able to keep manure far enough away from the horses who make it to live without the aroma in the background. The neighbors would also be far happier if manure were removed frequently, or never showed up at all. However, we have to take into account the time and energy we have. When it comes time to clear the paddock of a ten-horse herd, there's a huge difference between pushing wheelbarrows full of manure 1000 feet uphill through mud, or 100 feet or so over paved surfaces.

You can adjust the height of your shelter to suit the breed or breeds of horse you keep: for those who care for a variety of breeds and sizes, a ceiling height between 8 and 10 feet should be enough.

An open barn, airy and well-lit, with adjoining paddock and pastures. The horses have free access to one side, and people, the other.

We can choose a reasonable compromise: manure can be kept out of sight of neighbors, off public roads, and away from the edges of pastures. This will keep it from contaminating rainwater. Ideally, the field where manure is dumped will be surrounded by a trench to catch liquid waste seepage, but will also be close enough to avoid exhausting the person pushing the loaded wheelbarrow.

Efficient Use of Space and Walkways

If we're planning to provide hay inside our shelter, it makes sense to position the shelter so that we don't ruin the whole paddock with our trips to and from the hayloft. Even if this space is completely paved, having a short route to and fro for large loads will save time, reduce mess, and make the work easier.

If everything is under a roof, the dimensions of any entrances should leave enough space to let you enter easily with whatever equipment is needed for cleaning, feeding, or bedding the horses.

Consider Your Neighbors

The most important thing when keeping your horses at home is maintaining good relationships with your neighbors. Just as we wouldn't want to look out our window and see a wall rather than a pretty view, your neighbors won't like having the view from their house obstructed by a run-in shed. Give your neighbors as much consideration as possible when building something that will block the sun or cover up a pretty view from their window.

Water and Electricity

Whether water and electricity will be included in the initial construction, or you're planning for the future, plan to place your shelters as close as possible to the nearest hookup. Building far away from your existing wiring will make any future installation expensive and difficult when the time comes. Hooking up lights from a distance of one hundred feet is far easier than doing it from one thousand feet. Even if water and electricity aren't provided to the immediate area yet by your municipality, you should be able to request maps of planned networks and future connections. If you're near one of these places, take that into consideration when you're choosing a site for your shelter, so that any distances for potential excavation, cables, and pipes are as short as possible.

Encouraging the Horse to Seek Shelter

Build Where Your Horse Already Prefers to Be

This means a place from which your horse will have a good view of his surroundings and may also have natural protection from the elements. If at all possible, choose an area where your horses already prefer to rest or where they are used to having their hay. A perfect shelter in an inappropriate spot won't be of any use to horses who prefer not to use it. A friend of mine built a beautiful shelter, and her horses never used it because it was near a chicken coop behind a hedge. The horses didn't appreciate the comings and goings of their feathery neighbors, and although they adjusted to it in time, they only began to use the shelter regularly once my friend relocated it.

Feeding to Encourage Movement

Placing hay farther from the shelter in a basic enclosure should encourage your horses to move more, and they probably won't stay under cover as much. It will depend on what your needs are in the moment; if your horses are spending too much time in the shelter, place their feed as far away as possible. If it's always vacant, however, bring the hay closer.

Standing Up to the Wind

If, on a windy day, your shelter has a rattling roof, wobbly walls, and creaking planks, a sensible horse will make every effort to steer clear. Many herds kept in fields with tree cover will stand in the open with their backs to the wind rather than risk being under a tree with falling branches. They understand the safety risks associated with unstable shelter, and this is why so many shelters stand empty on windy and rainy days. If you want to build a quality windbreak and cover from the rain that your horses will be happy to use, it must be very stable and durable.

Dark Shelters Are Uninviting

Horses aren't drawn to dark, cell-like structures. Complaints about the stupidity of animals who avoid shelter in bad weather aren't justified—they aren't using their shelter because they're more comfortable out in the elements than they are inside it. Horses will absolutely use a well-placed, solid, airy, and spacious shelter, and will be relaxed and content; but if all they have is a dark, closed-in space they don't like being in, then they'll prefer standing out in the rain.

Frequently Asked Question:

Our horses are kept outdoors year-round. They have plenty of tree cover available to them, which they use to hide from the rain. Should we also build a shelter, or can we count on the forest to protect them?

You should build a shelter! They might like to stay under the tree cover sometimes, whether they have another shelter or not. But if you are keeping them outdoors year-round, it's your responsibility to provide a permanent, sturdy, comfortable shelter they can use, if they don't have direct access to their stable—tree cover or no tree cover.

Shelter Dimensions

There are minimum standards for the dimensions of stalls for the protection and welfare of individual horses, but these aren't open shelters—they're buildings where animals are enclosed. Minimum dimensions for an open shelter aren't defined in the same way.

In order for a shelter to be functional and inviting to your horses, you must keep in mind that horses are instinctively distrustful of small spaces. If you would like to be certain that whatever you build won't discourage the herd, it must be sufficiently open.

If a dangerous slope is created during the construction of your shelter, it can be leveled by pouring stone.

In this photo, it looks like a barn, but in reality, it's a spacious, open shelter.

Pasture shelters can be found at farming equipment centers.

Reinforcement of the enclosure footing with concrete pavers.

Table A: Minimum Stall Space for One Horse

Height of the horse at the withers in hands	Individual Stalls[1]		Stalls for foals and for a mare and foal[2]	
	area in square feet	shortest side in feet	area in square feet	shortest side in feet
< 8 hh	30	5	40	5.5
8.1 to 9.3 hh	45	5.5	50	6
10 to 12 hh	55	6	70	7.5
12.1 to 13 hh	65	7	80	8
13.1 to 14 hh	75	7	90	8.5
14.1 to 15.2 hh	85	8	105	9
15.3 to 16.2 hh	95	8	120	10
> 17 hh	110	9	140	10.5

Footnotes:

1. For temporary stabling, the area may be reduced to 85% of these dimensions.
2. A mare with a foal can be kept in this space until the foal is six months old. After that, consult the space requirements for group stabling situations.

Table B: Minimum Stall Space for One Horse Kept as Part of a Group

	Multiple stalls—area in square feet	Layup stall[1]—area in square feet
Adult horses 2 years and older	100% of the area for adult horses according to Table A	80% of the area for an adult horse according to Table A
Young horses: 13–24 months	75% of the area needed according to Table A for the horse's expected height at maturity	60% of the area needed according to Table A for the horse's expected height at maturity
Foals: 6–12 months	50% of the area needed according to Table A for the horse's expected height at maturity	40% of the area needed according to Table A for the horse's expected height at maturity

Footnote:

1. This means space available for lying down. Feeding equipment must not be included in this area. If the horses can move freely in the shelter and feed is also kept there, the same parameters apply as for group stabling.

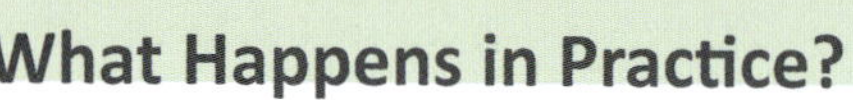

Shelter combined with hay storage in the rear.

Barriers above feeders, such as the ones pictured here, can rub horse's manes—take care, especially with animals with fine or brittle hair.

What Happens in Practice?

Everything. In some instances, horses will pack themselves into incredibly small, dark spaces for prolonged amounts of time, or won't use spacious and airy shelters to the extent that their owners would like. But one thing is certain: If we put enough effort into construction and make the shelter as large and as functional as possible, it will eventually become a sought-after refuge whenever the weather turns. Don't be discouraged by anyone who claims building a shelter is pointless—it's not true! If the structure is built well, the horses will like it. An added bonus: we won't have to worry about extra blankets for wet, miserable horses stuck out in the wind and rain, and we won't have to find solutions for coughing and eternally wet, cold-backed animals.

A beautifully designed layup stall within the paddock. The patient still has contact with the herd.

My horses (and the horses of my friends who have spacious shelters) flee their pastures for the comfort of their paddock and shelter at the first sign of rain. It's a huge comfort to know that I'll come home to dry, happy horses on even the most miserable days without having to rush home to put them in their stalls.

Area

It's a good rule of thumb to take into account the minimum recommended stall size for a medium-sized horse: 85 square feet. Multiply this minimum amount of space by the number of horses you'll need to shelter, and you'll have a good idea of the minimum size you need. Remember, if you can build larger, even better!

Height

The height from floor to ceiling of a stall (and this rule can also be applied to shelters) must be at least 1.5 times the height of the horse at the withers, with a minimum height of 7 feet. Some recommend 8 feet at the minimum, but I personally prefer a much higher height for a main shelter—which has the advantage of improving air circulation, too.

Material

There are countless options when it comes to building material: from wood, brick, or stone to garages made out of sheet metal (these become intolerable in the sun), metal structures with mesh (great for summer, but offering little protection from winter wind), or DIY (for example, with plastic

A makeshift attempt at a shelter, an example of wasted resources.

A pleasant stable, which horses will seek out year-round.

pallets—which can become brittle after a couple seasons in the sun, creating sharp fragments, so be careful!).

For the sake of both functionality and aesthetics, a shelter should be built properly the first time. Makeshift attempts that need to be replaced cost time and money and, most importantly, risk the safety of the horses.

Flooring

The minimum slope of the ground should be 1.5 percent. The easiest solution for flooring would be to leave it as dirt, but unfortunately, the easiest solution in this case is not the best. Over time, with regular use, dirt can turn into a hard, concrete-like surface. Or, with enough rain and snow, we can have the opposite problem: gusts of wind will carry precipitation into open shelters, and if the horses spend a lot of time there, soon the wet ground will become trampled and muddy, making it impossible to remove manure without taking some of the floor with it.

Another difficulty with dirt flooring arises if horses tend to pee in the shelter. Not all horses will, but if you get regular rain, one or two horses tend to pee inside, and you have more than just a few horses, I advise you to cover the shelter floor with something. Even if you don't mind regularly mucking out your shelter, a solid, flat surface will make cleaning and feeding far easier. Eventually, dirt floors become dimpled and uneven, and even the best efforts to flatten and fill them won't repair them.

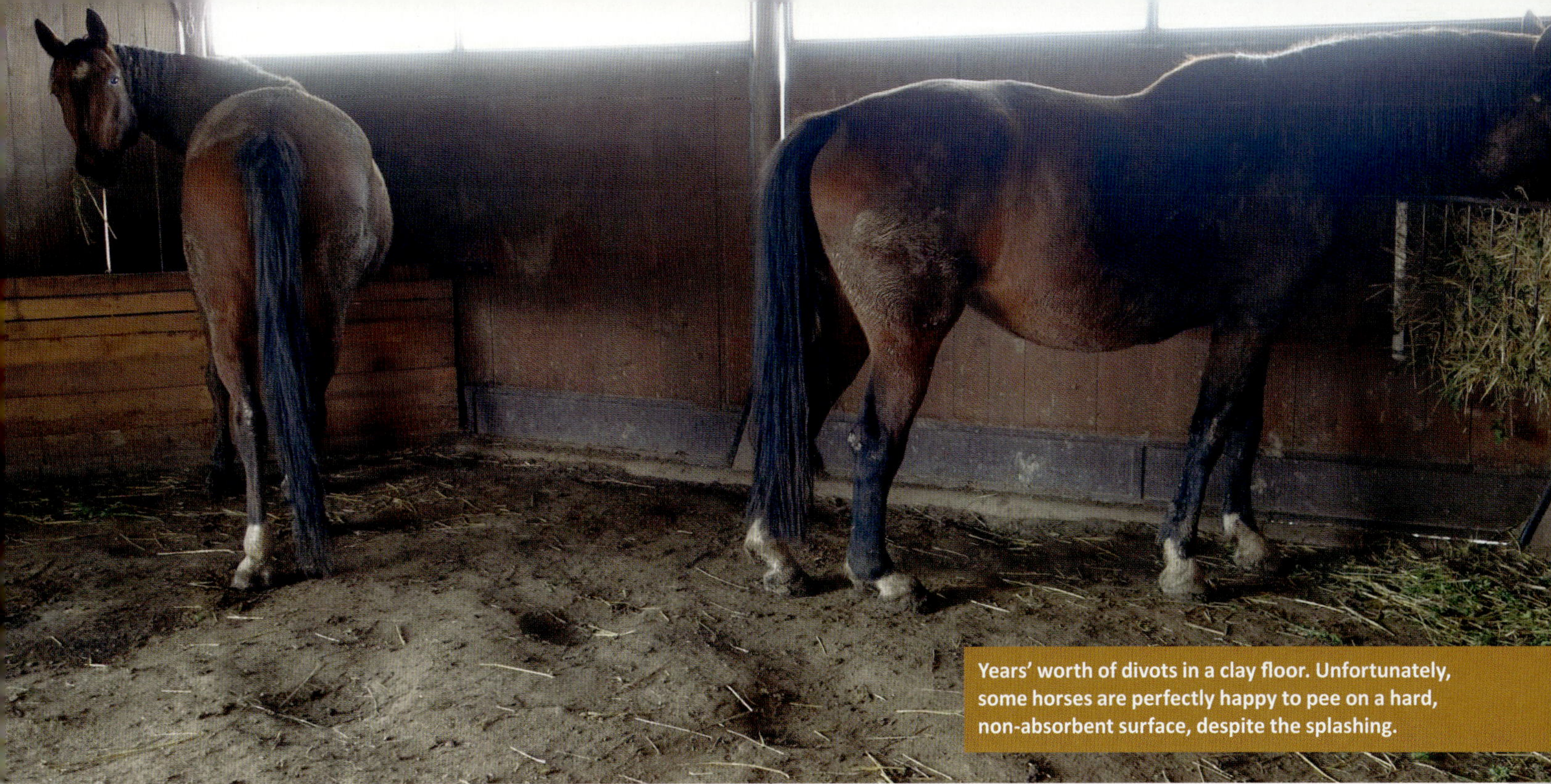

Years' worth of divots in a clay floor. Unfortunately, some horses are perfectly happy to pee on a hard, non-absorbent surface, despite the splashing.

Materials for Reinforcing Shelter Floors

Sawdust, Shavings, Straw, and Other Common Soft Materials

Many owners love to use these over dirt, saying it keeps the footing clean and compacted. Personally, I've only had success with this option in the short term, even when used in shelters for a small number of horses. When used day after day by the whole herd, even shavings or sawdust with a layer of straw over it can't prevent urine from penetrating the ground and causing all kinds of lumps and divots...

Plastic Grates, Tiles, and Mats

Various materials of different shapes and sizes, like rubber or plastic tiles, boards, or grates, can be added to a surface to reinforce it. Whichever one you use, if you want a final floor that is flat and stable, the surface underneath your extra material needs to be correctly prepared—you can't simply lay these on dirt and call it done. The surface should be leveled, covered in gravel, and thoroughly compacted. Otherwise, urine and blown rain will eventually ruin your flat surface, making it wavy and difficult to clean. Smooth mats are a popular option, but can become very slippery when wet. (Some can be slippery even when dry—I have had my own bad experience with this!) Mats with holes need a good base and a final fill with fine gravel or sand. These can work in any space horses use less frequently. Solid mats are the top layer—no additional fill is required. These can be used alone, or bedding can be added. It depends entirely on what

Mats laid directly over dirt, without gravel substrate.

Example of a correctly laid shelter floor made of solid mats over a gravel base.

system works for you, and whether you want your horses to lie down outside or under a roof.

If you decide to use perforated mats and use only a small amount of gravel or sand underneath them, especially if you then combine this with some kind of absorbent bedding (so your horses will be comfortable urinating in the shelter), the mats will start to ripple over time. The large amount of urine horses produce will eventually be absorbed by the soil under the gravel as well, creating an irreparably uneven floor. With this in mind, it makes far more sense to use solid mats for your floor, and to install them at a slight slope so that urine can drain away to a designated point. If you want to avoid an unstable foundation and the smell of urine under the mats, I strongly recommend not using bedding in your shelters, especially if you have a whole herd of horses!

If you set up a demarcated open-air space (a so-called "urination station"—see the next chapter) away from the shelter, even a small one... you'll find that most horses will choose to urinate there (although this will need to be taught—it's worth it to take the time to do so!). If you still prefer to use permeable mats in the shelter, and fully intend to have your horses spend plenty of time there, be sure to construct the foundations underneath those mats correctly: gravel will need to be layered carefully, from the coarsest size to the finest, with each individual layer fully compacted before adding the next.

Tip

If for some reason you can't add any additional materials to the soil, and you'll be using mats without any kind of fill underneath, I recommend using heavy-duty carpeting underneath your mats to at least prevent mud from squeezing up through them when your horses step on them.

Wood

The use of oak in building stables has a long history. This material creates a warm, long-lasting, firm yet flexible flooring. If you'll be adding bedding over a wood floor, oak is an excellent choice for its resistance to rot and mildew; some people like acacia for this as well. Although oak boards or planks are a good choice, do not use wooden railway ties. These are considered hazardous waste due to the chemicals used in treatment and production, and subsequent contamination by railway traffic.

Concrete panels surrounding a feed area are a low-maintenance option.

These shouldn't be used indoors, or in areas where animals will sleep or eat. However, they can be used to reinforce road surfaces, yards, and parking and storage areas for cars and heavy machinery. They can also be used for outdoor stairways.

Concrete

Good concrete has the benefit of being very durable, but there is a big drawback: it is cold. When concrete pavers are used as a shelter floor (or in a feed area), they can be quite slippery, especially when horses go from lying down to standing up. Therefore, many people choose to add a thick layer of sawdust or shavings on top, and others put straw on top of that.

Alternatively, some people use pellet bedding. Of course, with a bed like this, horses will choose this area to urinate and lie down. Personally, I prefer not to add bedding in shelters—we're keeping horses outdoors! There's nothing worse than a shelter that a horse won't leave because they have food, a bed, and a toilet all under one roof.

Even the toughest horses shouldn't run the risk of being trapped in their shelter. Always build several entrances (unless the entire front wall will be open).

A much better choice for concrete floors is to cover them with rubberized load-bearing mats, which allows the horses a comfortable place to rest and lie down in inclement weather without the need for bedding. Since the goal is a functional dry shelter, we'll create what I like to call a bathroom for your horse (further explanation in the chapter entitled "Urination Stations"). If your horses have already become comfortable peeing in their shelters thanks to having them bedded with some absorbent material or other, it might be difficult to re-train this behavior, even if we remove the bedding and place it outside, especially if it's a strong-willed animal. If you decide that you don't mind your horses using their shelter as if it were an open stall, spending more time than necessary in it, and doing all their business between its walls, then it will be necessary to keep the area well bedded so your horses won't slip on wet floors when rolling or getting up.

Bricks

If bricks are used as flooring anywhere under horse's hooves, we should be sure to use twice-fired bricks. These bricks are created using extremely high temperatures and are fired twice, making them quite durable. These bricks can withstand humidity, frost and sunlight far better than cheaper bricks made conventionally. When used as flooring in a structure for horses, they should be placed on the narrower edge rather than the wider surface, ensuring the durability of the final product. You can expect a brick floor made to these specifications to last a very long while, especially if your horses are barefoot.

Stone

If you have access to large, flat stones, they can make an excellent floor. I've seen a basic paddock paved entirely with stone, and it was perfect! There was no mud. The only unpaved space was the space provided as bedroom and bathroom: a rectangle covered with sawdust and straw. All the other surfaces were made of flat stone, here and there sprinkled with fine sand, and everywhere else cleanly swept. The horses' hooves were in amazing condition.

When laying stone, level the surface and prepare a sandy base. Try to keep the gaps between the stones as small as possible. The size of each slab must be such that they're anchored by their own weight, and also stabilized by their proximity to the stone next to them.

Gravel and Stone Dust

As a finished surface under a shelter, this is an unsuitable material. Gravel and stone dust will never create a smooth surface, and when you're mucking, the integrity of the footing will be ruined as bits of it are pulled out with the manure. When you're dealing with manure with gravel stuck to it, cleaning is slower, wheelbarrows become heavier, and rocks end up in the compost and therefore follow you into your garden beds and pastures. Small pieces of gravel may not matter, but larger stones in manure become annoying quickly. Horses themselves don't mind lying down on smaller pieces of gravel, but larger stones don't encourage horses to lie down to rest. If your horses don't lie down much in their shelters, then a gravel surface will bother them far less than it'll bother you, but I can't recommend it as a top layer in any kind of shelter.

In the summer months, it can be great to have entrances opposite each other to encourage air flow. In the winter, however, a windbreak is ideal. These opposing demands can be met by using movable walls which, when rearranged, fill the needs of the season at hand.

Roofing

The roof is a crucial part of construction, yet it often gets the least amount of attention. I've seen plenty of shelters that were never fully functional because of a poorly constructed roof! A run-in shelter is just like any other building, with many of the same requirements. It makes no sense to give up just as construction is reaching completion and "slap something on"...

The material you decide to use for the roof is of the utmost importance. If you use a reputable company to build your shelter, they'll give you good advice, but if you're building it yourself, you'll have to make your own decisions. I asked a roofing company in my area with years of experience for help, and based on their advice, I've assembled some recommendations for you here which can help you navigate this aspect of construction. In my experience, bituminous (from now on referred to as "tar paper") and corrugated metal roofing are the most common, but there are many other options as well. Some of the most-used basic roofing materials are:

- **Asphalt shingles**
- **Concrete tile**
- **Tar paper**
- **Aluminum**
- **Clay tile**
- **Sheet metal (steel)**
- **Fiber cement shakes**

Roof Slope

Clay tile—for a roof with a minimum slope of 22° and above.

Concrete tile—for a roof with a minimum slope of 17° and above.

Aluminum—for a roof with a minimum slope of 14° and above.

Fiber cement—for a roof with a minimum slope of 15° and above.

Asphalt shingles—for a roof with a minimum slope of 7° and above.

Sheet metal—for a roof with a minimum slope of 5° and above.

Tar paper—for a roof with a minimum slope of 5° and above.

Price comparison for basic roofing materials:

Clay tile	$15 USD/sq yard and up
Aluminum	$11 USD/sq yard and up
Concrete tile	$10 USD/sq yard and up
Sheet metal (steel)	$10 USD/sq yard and up
Fiber cement	$10 USD/sq yard and up
Asphalt shingles	$10 USD/sq yard and up
Tar paper	$8 USD/sq yard and up

Product life (according to manufacturers):

type:	service life:
Aluminum	minimum 50 years
Clay tile	minimum 50 years
Concrete tile	minimum 50 years
Corrugated metal	minimum 35 years
Tar paper	minimum 20 years
Fiber cement	minimum 15 years
Asphalt shingles	minimum 15 years

Comparison of roofing materials according to environmental impact:

Roofing Material	Energy						Overall Evaluation (see below)
	PET (kW/m²)	PET (see below)	CO_2eq (g/m2)	CO_2eq (see below)	SO_2eq (g/m2)	SO_2eq (see below)	
Wooden shingles	14	Level 3	-16,156	3	27.300	3	Level 3
Reeds	8.775	Level 3	-71,955	3	122.175	2	Level 2.67
Concrete tile	10.560	Level 2	6,336	2	22.080	3	Level 2.67
Slate	-	Level 3	-	2	-	2	Level 2.33
Clay tile	50	Level 2	17,500	2	61	2	Level 2
Synthetic slate	54.720	Level 2	19,036.8	2	75.744	2	Level 2
Asphalt shingles	70	Level 2	5,240	3	32.450	2	Level 2
Steel panels	123	Level 1	30,592.5	1	156.675	1	Level 1
Plastic panels	-	Level 1	-	1	-	1	Level 1
Aluminum panels	615.600	Level 0	113,297.4	0	820.800	0	Level 0.25

PET = polyethylene equivalent; CO_2eq = carbon dioxide equivalent; SO_2eq = sulfur dioxide equivalent

Environmental impact level:

Level 3: Highly recommended.

Level 2: Recommended, some minor drawbacks.

Level 1: Significant drawbacks, not recommended.

Level 0: Best to avoid.

Mass:

Type	Mass
Clay tile	9.3–12 lb/sq ft
Concrete tile	8–11 lb/sq ft (double wide: 14 lb/sq ft)
Fiber cement	2.4–4 lb/sq ft
Asphalt covering (⅛ in)	2 lb/sq ft
Steel panel roofing	1 lb/sq ft
Tar paper	.84 lb/sq ft

Practical Advice

Anchoring Loose Roofing

Tar paper will need twice as many nails as the manufacturer recommends to hold it in place on a windy farm. Also, using a washer under each nail, of the same diameter as the head of the nail or wider, is highly recommended. In windy conditions, this roofing will very quickly develop holes around each nail if smaller washers are used, and eventually a strong gust will simply rip the roofing off. In the meantime, water can easily leak through the holes. Without appropriate fasteners, this can start happening in as little as five years...

Metal roofing offers very low thermal resistance. When the outer surface of such a roof cools (during the night, or under snow), condensation can form on the underside and drip on the horses. The good news: there is already sheet metal on the market with anti-condensation treatment on the inside, made of interconnected fibers that hold the condensation in small chambers until warming temperatures allow it to evaporate.

Metal Roofing

In the winter, this kind of roofing will cause water to condense and drip on your horses if it doesn't have insulation. Some people don't mind, while others anticipate this problem during construction. It's an easily fixed problem—either purchase sheets designed to prevent condensation, or place insulation under them.

How to encourage your horses to use their new shelter:

There are many ways to make this new sheltered space seem enticing, some of them very easy to do.

Things to try:

- Hang hay nets indoors, at least until the horses stop being suspicious of it.
- Fasten branches to the walls or simply place them on the ground. Especially in winter, horses may want to find something to chew on, and may find strategically placed branches enticing.
- Feed your horses not only hay, but green grass and other tasty fodder in your new shelter. With larger groups of horses, focus on your lead gelding or boss mare: if you convince this horse that the new shelter is a safe, comfortable place, they'll probably lead the whole herd to it. It

bears repeating—make sure your shelter is the appropriate size for the number of horses that will be using it.

- Install a fan in the summer; it will drive away insects and significantly improve the temperature.
- Only clear the area immediately surrounding the shelter in times of heavy snow and shovel out the entrances. Don't be afraid to bed the shelter with straw or shavings. To really raise the standards, add buckets of lukewarm water alongside hay nets in the shelter. Of course, this pampering of your horses should be temporary; although you should use every trick to make your horses feel like this covered space is a VIP area when introducing them to it, once they're comfortable with it, you can move their food and "bathroom" outside.
- Make sure there is plenty of dry bedding inside when it rains, until your horses become accustomed to their shelter, even if you don't intend for them to use it as a place to lie down all the time.
- Hang salt licks and toys (even a plastic jug can be a toy if your horse isn't afraid of it). These will

A well-lit shelter, thanks to translucent roofing.

be irresistible for playful horses, and for the less playful ones, these objects will still pique their interest and inspire their curiosity.

- Save any interaction between you and your horses (like grooming, scratching, and pampering) that your horses particularly enjoy for time spent in the shelter.
- Imagine this shelter as a beautiful, convenient bonus space under which your horses will find nothing but comfort and happiness. If you believe this, you can turn your belief into reality, as your horses will notice this attitude and begin to share it with you.

If you have two horses that won't choose to enter the shelter on their own, and it's been designed to safely serve as a double stall, close them in at night while they're becoming accustomed to it. This will prevent them from running out each time they hear an unfamiliar "new construction" sound, which is not just counterproductive but can cause them to injure themselves.

Lightning Rods

These aren't mandatory, but it can be a good idea to install a lightning protection system on any structure that stands alone in an open space or on a hill, especially if it houses a large number of animals.

Did you know? Storms provide natural fertilization. During thunderstorms, electric atmospheric nitrogen oxidation occurs. This reacts with rain water to form nitric acid. Nitrates formed by this reaction are processed by microorganisms in the soil, creating substances that fertilize soil and feed plants.

Bedding or No Bedding?

Dangerous Gasses

The air in stables contains more water vapor, CO_2, and microbes than the outdoor environment. This has a significant impact on horse health. Outdoor shelters are no exception, if they're even partially enclosed. Manure, when mixed with bedding, can form harmful gasses. These smelly gasses contain nitrogen, oxygen, and sulfur. When combined, these gasses intensify each other. They not only damage the health of humans and horses, they also affect the environment. Eliminating odors isn't just a matter of avoiding smelliness, it's a matter of animal welfare. Most smelly gasses amplify each other—for example, ammonia, with its characteristic pungent odor, makes everything smell worse.

Listed here are the most common gasses found in a stable:

- ammonia
- methane
- CO_2 (carbon dioxide)
- hydrogen sulfide

In addition to these primary health-threatening gasses found in our stables and shelters, additional gasses are produced that can be harmful even in small amounts when commingled. These gasses, individually negligible, can have a significant impact on a horse's health when combined. For example:

- indole
- butyric acid
- mercaptan
- skatole

Various odor eliminators for rot and mold are available. The best types are those that don't just cover up the odor, but absorb it. I recommend focusing on environmentally friendly materials so they can be used everywhere, including areas where your horses prefer to poop and pee, and also where they lie down. Of course, we won't be constantly mucking and changing bedding like we would for a stalled horse, but to keep the concentration of urine at an acceptable level, areas with bedding should be picked every few days or, depending on the number of horses, completely changed. However, if your horses decide to pee directly on soil, it can become quite soaked. Depending on the degree of odor, we can toss a few shovels full of sand over it, or use one of the various odor-neutralizing products available for agricultural use. Keep in mind that if you're using an environmentally friendly product, it can also be used inside the shelter to keep intolerable odors at bay.

Dust

Another issue that can be especially suffocating in stables (or in our case, shelters) and needs to be dealt with is dust. Dust particles, combined with the air in an often-used shelter, can be quite harmful. Studies focusing on this have concluded that dust is, unfortunately, ever-present at harmful levels in stables, and damages the airways and mucous membranes of stalled animals. Furthermore, dust can carry bacteria and microorganisms. The amount of dust in the air increases proportionally with the number of animals. The amount and composition of dust particles depends on the type of shelter, climate, season, use of bedding, and, last but not least, the daily care of the shelter and the entire basic enclosure.

The term **"biological aggressiveness,"** when discussing dust particles, refers to how they enter and affect the body. Larger particles are gradually brought to the digestive tract, where they're metabolized. Smaller particles are captured by ciliated epithelium in the mucous membranes.

A horse's respiratory system is not well-adapted to a dusty environment, whether a stable, shelter, or

Average dust in an agricultural setting contains a wide spectrum of elements, from plant fragments, bacteria, mites, fungal microorganisms, allergens, and insect particles up to insecticide contamination.

simply a dusty or moldy feed tub. The most dangerous particles are the smallest ones (with a diameter or less than 2.5 micrometers) as 90% of these end up being trapped in the pulmonary epithelium—that is, in the horse's lungs.

- **Biologically inert dust particles** (which are inactive and have no specific biological impact) have an effect, but the consequences for the organism are reversible. The body can "clean the lungs out," removing this kind of dust from the mucous membranes by coughing. If the body's limits aren't exceeded, coughing will clear the lungs and airways just fine.
- **Biologically aggressive dust particles** ("active" dust) can cause extensive health problems. Depending on the composition of this dust, the effects on a horse's respiratory system can be severe. If the composition of the air includes approximately 10 percent "active" dust, prolonged exposure can lead to chronic problems.

Dust from silica, although inert, can be problematic. In construction settings, for example, silica is the second-greatest risk to workers after asbestos. Because its particles are very fine, it can become trapped in lung tissue, which responds by forming fibrous nodules and scarring. Long-term exposure can cause lung cancer and other serious respiratory diseases. According to the International Agency for Research on Cancer, silica dust is classified as a group 1 carcinogen—a clear cause of cancer in humans.

When you are considering what type of sand to use for the basic enclosure, for these reasons, I personally don't recommend using silica sand.

Frequently Asked Question:
How is it possible that those who work with glass or in construction are often diagnosed with silicosis, but people who live in desert environments don't suffer from this condition?

The answer to this question can be found in the physical properties of dust particles. Dust from glass, for example, when looked at under a microscope, has sharp edges. Dust from sand in the desert, on the other hand, is round. This smooth shape doesn't irritate airways, even though it's also primarily silica dust.

Sharp, needle-shaped particles most easily penetrate epithelial cells of the lower respiratory tract. They literally bury themselves and can eventually cause everything from respiratory infections to cancer. Round particles are much less dangerous.

In this chapter, I have addressed what I consider to be the biggest risks associated with a dusty space; there simply isn't room in this book to address them all. In the context of keeping horses outdoors, I can't stress enough how important it is to reduce the amount of dust in your horse's environment as much as possible. If at all possible, it's best for horses to live on healthy, undamaged (and therefore dust-free) grass sod.

The worst areas for dust are enclosed areas of stables, but outdoor shelters can also become dusty, especially if horses are fed there. The risk of developing respiratory issues is on the rise, which is why many people who keep their horses outside refuse to lay bedding in their run-ins. Although it can cost more in materials and labor, if you choose to use bedding, use it outside of the shelter.

When discussing dust, I'd be remiss if I didn't address clay paddocks, which in a dry climate can literally form a layer of dust under the horses' hooves. This could be solved with regular watering, but that kind of endless use of water isn't a very environmentally friendly approach. Instead, we can address this issue by using a dustless options for footing. (See chapter 4 for more on reinforcing the surfaces of the basic paddock.)

Through correct ventilation, sufficient shade, repairs to dusty surfaces, frequent removal of manure, sweeping only when the animals are elsewhere, and limiting the amount of bedding used, we can optimize the atmosphere for our horses in their shelters.

chapter IV.

REINFORCING FOOTING

It's my strong belief that creating a durable and healthy paddock for horses requires reinforcement of at least some part of the surface.

No matter how large a paddock or pasture is, there will always be an area the animals prefer to use regularly. This constant use will result in mud as soon as there's steady rain. How deep or uncomfortable this mud will be will depend on location, climate, soil composition, and the size of the space. Also important: the number of horses and how they use the area. The larger the space, the less deep the mud, but only in the ideal situation where horses use the space evenly. Even if this is the case for your horses most of the time, it's unrealistic to expect it to continue in winter. Even horses who have year-round access to pasture will spend a large portion of the colder months gathered near shelters and feeders. Unfortunately, this can lead to swampy conditions...

The last couple years have shown most of us how unpredictable the weather can be. There are few regions where snow and freezing conditions can be expected throughout the entire winter. Warmer winters with multiple thaws have made a lot of horse owners consider reinforcing at least some parts of their outdoor spaces. Unfortunately, for many, the expense can make this a difficult task.

Is reinforcing footing really such an expensive undertaking?

Yes. If we want to make these improvements in a way that will have a long-lasting impact, we have to prepare to spend a good deal on the right materials. If you have the ability to shop around, some paving, load-bearing slabs, or curbs to demarcate areas may be found at building supply outlets for lower prices than from conventional suppliers.

Some horses love to roll in mud, but we want to be sure it's true love and not the lack of other options that encourages this messy behavior. We must give our animals enough dry space so they can always escape the mud if they wish. A horse rolling in mud or stumbling through muddy ruts when there is no other choice isn't a mud-loving horse, but a suffering animal. Many animal welfare groups consider conditions such as these worthy of a fine.

In practice, you'll see a difference almost immediately once you've reinforced footing to create dry areas. Horses will voluntarily stay on the drier, reinforced ground, and will go into the wet or muddy areas only occasionally.

Living in a situation like this one, without any way to find relief on a dry surface nearby, is not ideal. The fact that some horses can stay healthy in such conditions doesn't mean that it's a good idea to leave things like this. After this photo was taken, this area was reinforced with recycled brick, and is now in much better condition.

And what about us, their owners and caretakers? Of course we hate the mud! If we get stuck in a paddock once because the mud swallows our boots, it's a funny story. But if it happens every day, it gets less funny very quickly. There's no reason other than limited finances to put up with a situation like this.

If you're going to take responsibility for your horses' comfort and well-being, you should plan so you're able to invest as needed! (Ever remember being told something like this about that pet you wanted when you were young? "It's not a free [kitten, puppy, hamster...], you're signing up for all the care that animal will need.") Our outdoor horses don't go out for a few hours before being taken back to their stall. They're in their pastures 24/7. If your horses aren't used regularly for work, there will be times when they don't get taken out of their enclosure at all. If you wouldn't want to be forced to stay in a situation like the one in the photo on the previous page, why would you ask your horses to?

Is it possible to beat the mud?

The good news: it is! The bad news: it's going to cost us—not just money, but also time and hard work. Those of us who don't own the land we keep our horses on don't have as many options for reinforcing the ground. Even people who do own their own land aren't necessarily excited to fill quality agricultural land with stones. Most farmers won't want to interfere with their own land in a way that is irreversible, and will seek out temporary solutions instead, ones that won't degrade soil. No one is going to live forever; what if our children and grandchildren want to have a vegetable garden instead of a paddock? They won't particularly enjoy dealing with brick or asphalt that's been buried by years of compaction by horse hooves, so a more temporary option may be advisable...

Temporary Landscaping

Options That Are Easy to Undo

1. Plastic Mats/Load Boards

There are several varieties of plastic mats and load boards on the market today, from various manufacturers. Some domestic companies can offer excellent prices on recycled PVC mats. Be sure that their PCB content level has been certified. These mats are usually non-porous and wear-resistant, and may be shaped so they can lock together. Some may be perforated, and will therefore be lighter.

Prices can vary from very low (hard to find) to many times more expensive (but easier to get). Although these mats are recommended for use in covered areas, many owners have used them successfully outside as well.

Solid PVC mats, although designed for indoor use in stables, have been used outdoors with great success by plenty of horse owners.

The clear difference between a reinforced section of a basic enclosure and the unaltered section. Horses will choose the dry surface more often than not.

A little wisdom based on others' experiences:

- Don't lay mats directly on unaltered soil. If the mat you choose is perforated, it will eventually sink completely into the mud underneath it. With solid boards, mud will form at the edges where the boards touch each other during rain. In these areas, divots will gradually form, trapping moisture and creating soft spots where horses can break through. If you prefer not to use several layers of aggregate, geotextiles can be used instead, backfilled with a smaller layer of gravel. Another option is to place heavy-duty carpet over the entire area to be laid with mats.
- The more areas you can reinforce, and the larger the mudless surface will be, the less load each individual board will have to carry. I myself leveled a large area to serve as a path to the horses' drinking water. In my experience, it's worth it to make this area as large as possible. If you've got just a few boards over a sea of mud, mud will squeeze through anywhere it can whenever the horses walk on them, requiring daily cleanup.

2. Rubber Tile or Slabs

You may have seen these used at a playground, or as a safety measure at a sports ground. With horses, they're wonderful, around feeders and under shelters—wherever horses spend a lot of time. Their smooth surface (unlike mats with bumps in their design) makes them a good option for spaces where horses lie down, and makes them easier to clean (since a shovel won't catch on the surface). Best of all, these mats aren't slippery, even after it rains!

Unfortunately, these are quite expensive, at over $40 for ten square feet. It's worth keeping an eye out for discounts at outlets and warehouses.

These rubber-puzzle squares were purchased at a more affordable price than usual at an outlet. The substrate used here was fine gravel.

3. Sod Mats for High Loads

Sod mats can be unrolled onto almost any surface without modifications, as long as the surface is reasonably level. They can be attached with garden staples which are about 6 inches long. Manufacturers often provide installation videos on their websites that are short and easy to understand. On one site, I found sod mats advertised as "solutions for creating a thick lawn without the need for any surface preparation, and at almost any time of year." However, if the area where you want to install a sod mat is already damaged, the same rule applies as for plastic mats: on clay, without any backing or aggregate, the mat will eventually sink into the mud.

Be sure to choose "high load" mats; these can stand up to horses. They're also suitable for creating a riding ring. Prices will vary depending on the size of the roll; for example, you may be able to find a 30-foot roll for about $650.

4. Interlocking Plastic Grass Pavers

I see these used quite often, and most horse owners I talk to are very happy with them. Essentially, these are plastic screens with holes that can be laid either directly on a level surface, or over a drainage cloth. Again, without a solid foundation, mud will pass through the holes, causing the pavers to settle and then undulate over time. After the base is prepared, these pavers provide erosion control and stabilization; cover the original screens with sandy soil and grass seed, and over time, as it grows, the grass will anchor the screens to the base. I recommend this only for little-used paddocks where you want to establish or reestablish grass. These pavers, like sod mats, come in various strengths and thicknesses. Be sure to get the option best suited for use with horses. Most suppliers will be happy to steer you in the right direction, as horse owners are some of their best customers!

5. Heavy-Duty Carpets

Carpeting of paddocks has become very popular in the past few years (perhaps by necessity?)—but mostly with owners of only a few horses. Commercial outdoor stabling has resisted this trend, mainly because under the hooves of a large herd, carpets will disappear into the mud. Digging out a large piece of carpet is a real treat! (Keep this in mind before you decide on this method.)

However, they serve their purpose well in a smaller setting. Many owners of small backyard barns can't praise them enough, even after several seasons.

Cons of carpeting (may not apply to smaller setups)

- Carpets tend to curl eventually, even when anchored with nails.
- If the horses urinate on them, eventually they'll smell quite bad.
- After a few seasons, they begin to tear, making it necessary to layer new pieces over the old.
- When using carpeting with multiple horses, there's a real risk of injury, especially with ordinary residential carpets (as opposed to commercial or industrial carpeting). They easily tear, and once they tear, a horse can step through the hole and catch his leg. Even correctly laid, load-bearing carpet under a group of quiet horses can cause injuries under the wrong circumstances.
- Even heavy-duty carpeting won't withstand everything. The same carpet will have a different lifespan with two or three older, calmer horses as opposed to a herd of youthful (and active) horses.
- The need to anchor carpeting carries risks. Personally, I'm very afraid to use nails anywhere around horses. Usually, carpeting is held in place with long nails driven into a pad, so the nailhead doesn't go through the carpet and disappear into the mud. Unfortunately, without them, no carpet can be held in place. The constant motion of horses can cause a carpet to roll up in the blink of an eye. Even a strong wind can roll a heavy carpet if it isn't anchored properly. I recommend using plastic plate dowels instead of nails; these can be pounded easily into softer soil after rainfall.

In some countries, especially in Germany, many owners of outdoor farms like to use artificial turf as an alternative to heavy-duty carpet, taking advantage of old rolls that have been removed from sports fields and are headed for a landfill. This material is significantly heavier and more durable than carpet, and can be used on solid surfaces without any anchoring. It's important to note that these mats are so heavy that they can't be moved or installed without machinery. The photo on the facing page is from a German stable, where a large herd of horses has regular access to turf, while pastures are closed off during rainy periods. Some owners choose to cover the turf with sand, while others use it as a base under load-bearing boards so mud doesn't rise through the gaps.

6. Concrete Grass Pavers

This method falls between the two categories "can be used on leased land" and "unsuitable for leased land." Whether these are reversible depends on how

Some stables in Germany and England use artificial turf that is being replaced and discarded. Often, under these circumstances, stables are only charged for delivery. This turf works very well when used over a leveled and backfilled surface.

they are laid—whether they go directly on dirt or in a gravel base. The advantage of concrete grass pavers is their indestructibility (as long as the correct thickness is chosen for use with horses).

Installation is also uncomplicated, although landscaping may be required. A quality gravel foundation layer is recommended, to prevent ice from pushing the blocks up into an uneven, wavy surface. A 6-inch-deep layer of 1–1¼-inch grit gravel should be sufficient for this purpose; this size of gravel should guarantee that freezing water will have space to expand. Then, a layer of sand is added onto the installed pavers. Finally, it is covered with fine gravel. However, I also know farms where pavers are installed directly onto the soil, or with minimal layers of sand, and last for years without any problems.

7. Concrete Flooring

Concrete tiles with smaller dimensions—12 x 12, 20 x 20, or 16 x 24 inches—when laid only on clay or in a layer of sand, tend to lift and move under horses' hooves, due to their smaller size and lighter weight. In most places I'm aware of where these have been tried in the paddocks, they didn't last. It's possible that they can be used more successfully in an area where horses don't tend to spend much time, or in a space that houses a pony or older, calmer horse. Older, used tiles can be found at great prices, and often can be gotten for free if you're willing to pick them up yourself.

8. Precast Concrete Panels

These panels are used in road repair, and come in different thicknesses and shapes. The land under them only needs to be leveled—no backfill required. The downside: these are extremely heavy and can't be installed without heavy machinery. Because it is such a hard material, and because of the risk of horses slipping and injuring themselves, many owners are understandably hesitant about using these panels.

Personally, I know people who love concrete surfaces for paddocks, and people who hate them. Depending on the person you speak with, you may hear only good things, or only bad... Road panels can cost as little as the price of delivery, or as much as $25 for ten square feet.

9. Wooden Railway Ties

These were used often a few decades ago, when there was less importance placed on the environment and safety. Today we know that it's dangerous to use these with livestock or gardening—in the US, it's outright illegal to use old railroad ties in home landscaping, due to the carcinogens that were used as the main preservatives. I discussed this in more detail in chapter 3.

10. Bricks

Bricks are a good option if your soil isn't clay and isn't extremely muddy. You may be able to find quality used bricks for the cost of delivery (as discussed in chapter 3, twice-fired bricks are recommended). If you have enough of them, you can try laying them on the narrow side instead of the wide side in order to increase their load-bearing capacity. You won't need a gravel base, just a layer of sand over a rough surface. Laid like this, they can withstand the wear animals place on them for a long time. Owners of barns with brick floors and

Concrete ties.

paths swear by them. However, if you don't use twice-fired bricks, you won't have the same good results. Ordinary bricks used in a wet environment with livestock will crumble quickly, especially if horses urinate on them.

11. Wood

Acacia or oak logs can be used for years, but the prices of these types of high-quality wood can be quite high. They are only a good option if you can install good drainage to avoid rot.

12. Sawdust, Shavings, Straw, Mulch, or Wood Chips

Many of us have tried this to cover mud, haven't we? For me, this option didn't work very well. I used to dump a thick layer of sawdust and shavings in a large area I used for riding, which the horses had free access to. It was nice in the summer heat and the horses enjoyed rolling in it, but after a few heavy rains, the shavings disappeared into the mud.

A straw-covered outdoor area is demanding. It requires frequent maintenance and replacement, making it expensive and wasteful.

Straw can keep mud at bay for a while if it's constantly added to. It can make a decent area eventually that can hold horses well enough, and if you use enough straw and keep adding more, not even rain will catch you off guard. But it doesn't provide a clean, dry surface—straw gets wet quickly and can't be relied on long-term. Last-minute cleaning, even with the right tools, can be difficult, and when you remove old straw, the earth underneath goes with it, which most landowners won't want. However, a combination of dirt and straw can be composted with manure to fertilize pastures. If the missing soil in the paddock is then replaced with sand, the problem is solved without sacrificing good stewardship of your land.

You can also wait for the rotting straw, manure, and urine to combine with the soil. Eventually, it will—but it won't be a solid mud-free surface, and won't be suitable for a basic paddock. Also, this process can take several seasons. For at least a few years, this area won't dry well and, when compared to the rest of the area, will seem (if I can be frank) really gross.

When the ground is frozen, a bale of straw can be spread to serve as temporary ground cover. For example, it can help create a clear, smooth path to a water source that won't be slippery. And a small amount of straw can be removed fairly easily when the time comes.

Some people who keep their horses outdoors swear by wood chips—and they need to be wood chips, not bark. They add more regularly, and have had no problems with mud. But according to others, wood chips decompose quickly and turn into an ugly soup that has to be removed.

13. Rubber Mats

I used to be able to find these for free if I agreed to pick them up, but these days they're more easily found for sale. These mats are used primarily in stalls under bedding, but many owners of just a few horses will put them in the winter paddock, too. I think they become slippery far too easily, but lots of people who use them seem to be satisfied, especially when they have over them a fine layer of dirt, sand, or sawdust...

Permanent Reinforcement

Options That Make It Difficult to Return the Space to Its Original Condition

1. Recycled Material as Backfill

- recycled concrete
- recycled brick
- recycled asphalt

Choose recycled materials that have been declared safe, and that meet the criteria of certified construction products according to any applicable standards!

Prices for recycled concrete, asphalt, and brick can be extremely low, but keep in mind that when the time comes to remove tons of material from your land, it will be very costly. The only way to truly clean such an area is by dredging and adding backfill. The good soil that you'll need for backfill can be hard to find, expensive to obtain, and expensive to transport. Check availability in your area.

Especially on softer soils, horses drive whatever reinforcing material you choose deep into the soil, meaning you have to add more and more as time goes on. It may be necessary to remove all that material, which is an intensive, time-consuming, and expensive process.

Before recycled material settles fully, it can be difficult to collect manure. Plus, loose material will be scooped up along with manure, which is a combination you won't want in your gardens or fields. If selling your manure, buyers won't want this mixture either. But after the surface settles and firms up, you'll be excited about how solid and consistent it is. Until, that is, the mud starts to seep through again, and the next layer of material is needed, and the cycle begins again...

2. Gravel as Backfill

If you don't mind adding stone to your land, gravel, pebbles, or other stones in some proportion can be added in at least some places. For example, this can be useful by heavily used pasture gates, in front of shelters, or for paths. Xenophon said it and it still holds true: if you want to strengthen your horse's hooves, stable flooring shouldn't be smooth or damp—it should be stone.

People who have a bad experience with horses moving over stone—frequent abscesses, for example—never return to using it, but those who find their horses' hooves are durable and strong due to walking over stone surfaces will sing its praises. For my part, I like busy places like feeding stations and popular rest areas (you'll know these spots by the large amounts of manure) to be reinforced in such a way that they're easy to clean; I don't want to end up picking up stones along with manure. I want my pitchfork to slide over the surface without grabbing anything unnecessary, which is impossible with a surface of loose stone. Then again, dealing with loose stone is far better than mining for poop in deep mud... It depends on your comfort level and your tolerance threshold. Although stone in these spots can at first seem like it makes for much easier work and improves the horse's welfare, I think you'll become more demanding over time, as you gain more experience and have time to try different materials and options.

I love using stone—for reasons both aesthetic and functional. The only difficulties arise with manure clean-up. So I like stone the best in any area where horses tend to pass through without lingering, since they rarely leave manure in these places. In areas that horses use more, I prefer other options.

How to Make a Mud-Free Stone Pathway

In order to create a resilient road, corridor, trail, or other stone surface, the gravel base rule applies. Remove the topsoil to a depth of at least one foot and thoroughly compact the area. Once it is compacted, place a permeable layer of drainage film to separate the soil from the stones. On top of this, lay the coarsest layer of stone and compact it. This layer, after compaction, should be about 7–8 inches thick. Use a slightly smaller size of stone for the middle layer, and compact this layer to a thickness of about 4 inches. This is topped with the finest layer of stone, which is also compacted. From time to time, this top layer will need to be refilled as horses carry bits of it off in their hooves and dig it up.

It's important to make sure that this path or area, when finished, sits slightly higher than the surrounding terrain, if you want rainwater to run off. However, if we want horses to have a "pond" to pass through during summer months, a stone path can be laid slightly below the level of the surrounding terrain, and the film used to separate stone from soil can be impermeable. During colder months, this area can be drained and left dry. In order to avoid having to dig up and replace the film each year, the stone layer directly above it should be made of smooth pebbles rather than sharp-edged gravel or stone. The film can be further protected with an additional layer of either rubber grids or plates between it and the pebbles.

Recycled brick can contain all sorts of things, even glass or nails. Be careful if you're using it as a crushed surface for your horses to walk on!

Coarse gravel.

3. Treating Surfaces With Concrete

Anyone who has experience with concrete and horses will tell you: invest in quality! Nothing is worse than having to repair and replace crumbling concrete. If you decide to lay concrete slabs in parts of your winter paddock, you'll need to remove the topsoil. The area will need proper drainage—either a sand or gravel bed. Especially if this area will need to withstand tractors or other heavy machinery, don't skimp on this step. The recommended depth of gravel in this case is 6 inches, and it should be compacted well. A frog compactor, which can be used in smaller areas, would serve well for this purpose.

The quality of the concrete used will depend on whether the area will only need to hold up to the weight and movement of horses, or if you'll also go over it with machinery (make sure you're sure—most horse owners will at least occasionally cross over paved areas with something like a gator or tractor!).

Blended cement is divided into several categories based on the ratio of cement to sand or stone. The good news: for use under horses alone, even the cheapest option should be sufficient. A talented mason can dare to use a larger percentage of sand and stones; even this can be turned into a smooth surface if enough care is used. Or, conversely, a trowel can be used to create ridges and grooves to prevent horses from slipping during winter months.

However, keep in mind—I can't emphasize this enough!—that if any machinery will be crossing over a given area, the concrete you use must be of a much higher quality. Even the pouring itself has stricter rules. If you'll be reinforcing a larger area, the space must be divided into smaller areas (approximately 6 or 7 square feet each) to prevent cracking in winter. These smaller sections should be separated by expansion joints, preferably polystyrene boards. The resulting squares will need to be reinforced with steel—mesh is a popular choice for our needs. The recommended thickness for the concrete slab is 4 or 5 inches to support horses, or about 6 to withstand machinery. This poured on-site concrete is what is called non-vibrated, meaning it will have air pockets and a weaker structure. This brings with it one large limitation: salt can damage it and cause it to crumble. After concrete is poured, it can be strengthened with a vibrating plate or rod, making it possible to salt without any worries—but this will add to the cost.

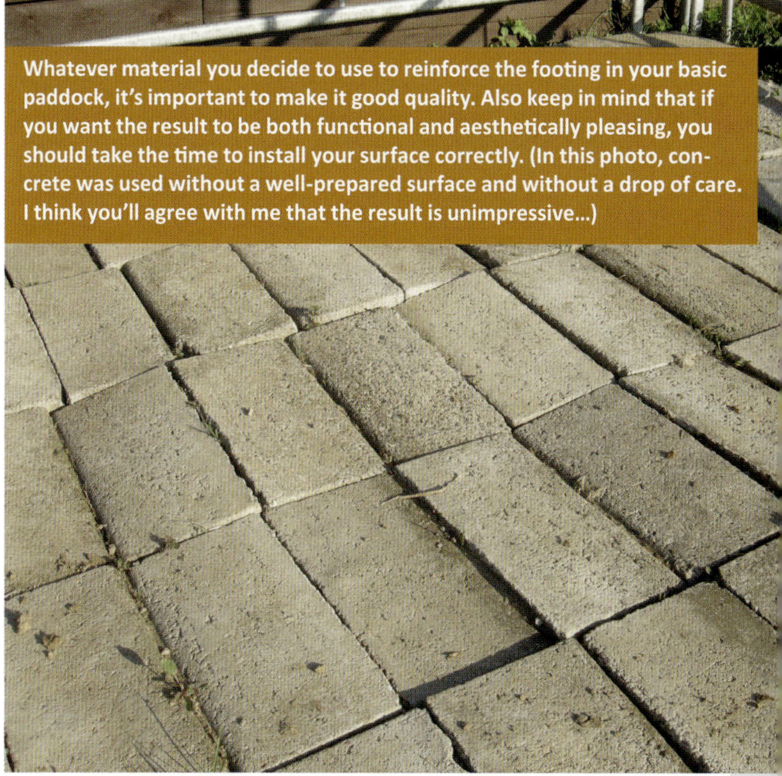

Whatever material you decide to use to reinforce the footing in your basic paddock, it's important to make it good quality. Also keep in mind that if you want the result to be both functional and aesthetically pleasing, you should take the time to install your surface correctly. (In this photo, concrete was used without a well-prepared surface and without a drop of care. I think you'll agree with me that the result is unimpressive...)

chapter V.

WATER

Figuring out how to keep horses hydrated is one of the fundamental hurdles of successfully keeping horses outdoors.

Especially in winter, many small farm owners struggle to find a solution that works well with their goal of allowing their horses unlimited time outdoors all year long. Carrying a couple buckets of water a few feet from the house to the fence line is no great feat. But for anyone who has to provide water for a larger herd, kept in a far-away pasture, when temperatures are plummeting, it's worth considering the level of work required and long-term feasibility. Years fly by, and an older body will have a harder time lugging water...

The horse's hypothalamus plays a critical role in regulating thirst. The feeling of thirst is triggered by changes to sodium content in the blood.

Having to push gallons of water in a wheelbarrow or carry it in canisters, while also having to navigate mud, frost, or snow, isn't the most fun! Think about having to do this on a day when your back hurts, it's very hot or cold, or you have a sprained ankle... and you have no one to help you. You still need to be sure your horses have water, at least twice per day. Imagine your morning starting like this: finally getting all this water to the pastures, knocking ice off the buckets so water can be poured—and you'll be back again in the evening to do it again. Some people can probably endure this, but I think it's worthwhile to come up with a better alternative...

Getting heated buckets that can provide horses with all-day and overnight access to water is, in my opinion, one of the best things you can do for yourself when building your horses' outdoor housing. If you place your basic paddock close enough to your home, there shouldn't be a problem connecting a water supply from the house. Then, horses can come to the waterer at any time from the pastures that are connected to the paddock. If there are spaces that are too far from existing water supplies, I would recommend drilling.

Even if you don't have the option to connect your water pump to electricity and have to buy a mechanical one in order to pump water from your well, it's still so much easier than bringing water from distant

The center for thirst is affected by the nervous system as well as psychological stimuli.

If at all possible, get your herd a heated automatic waterer instead of killing yourself carrying water to them. It's one of the best investments you can make in your outdoor horse housing!

sources, especially if you plan to keep horses over the long term.

There are also mechanical waterers into which animals can pump water themselves, using their muzzle or hooves, without the need for electricity. Although cows don't seem to have trouble learning how to use this unconventional system, the only reports I've heard of horses learning to do this are from people who also have cows in the same pasture. Some horses will manage to do it; others won't. Generally speaking, it's important for horses to drink well, as often as they need to, so the easier we make it for them to do that, the better.

One large horse (depending on weight, season, and workload) will need approximately 5–15 gallons of water per day.

An 1,100 lb horse in full work in a hot environment can drink up to 25 gallons per day!

Lactating mares can also drink up to 20 gallons or more per day.

Drilling for a Well

Being able to use a well as a source of drinking water is one of the first things that many landowners with a space without good access to water will think of. A new well, however, isn't a sure thing, as it may affect water levels in neighboring wells. You'll need to get approval and a permit for construction (a process which may include a hydrogeological assessment), as well as a permit for water management.

Once your well is completed, make sure your water use is legal; your local water authority will need to be notified that construction has been completed, and they'll need to issue you a permit. Until then, you can't legally draw from your well!

A drilled well accesses water at greater depths—where water exists almost independently of the amount of precipitation, and provides a more stable supply. This water is usually of better quality than that from shallower-dug wells.

The amount of water a well will yield depends not only on the geological conditions, but also on correct design for the well itself. So be sure to check references for drilling companies or excavators.

In some cases, there may be a generous buffer zone required, which may mean that you'll be told that you won't be able to drill for a well in the ideal location that you've found (for example, after drilling for an exploratory well) when the buffer zone is measured. But don't despair! In some instances, it may be possible to request an exemption. The paperwork can take a long time, but don't lose hope. No matter how long the delay may be, it'll be worth the trouble when you have your own well on your own land. Not only does it add value to your land, but it also adds so much for you and your horses.

The location of your well will depend on required distances from sources of possible pollution such as cesspools, sewer connections, barns, manure pits, tanks for liquid fuels, public communication lines, car washes... and so on. The location will also depend on the environment around the well.

Automatic Waterers

Each of these waterers is also available in a frost-free design.

Types of waterers suitable for horses:

- Level/float
- Paddle
- Nipple (pin)
- Ball

Where you don't have another option, you must use muscles—either yours or your horses'.

Ball waterers are freeze-proof. Some horse owners recommend removing the balls in the summer so the animals can drink more comfortably, but others leave them in, as they keep the water cooler. Dealers can discuss the advantages and disadvantages of each type of automatic waterer with you.

Think It Through: Automatic Waterers Aren't Always a Good Idea

With age, horses become less thirsty (probably due to a reduction in blood flow to the hypothalamus, where the osmoreceptors that stimulate thirst are located).

Many senior horses take in less fluid than they need, with consequences including increased blood pressure and deterioration of their ability to flush toxins. An attentive owner will notice some physical change in the horse and hopefully catch the problem in time, but the possibility of kidney failure in older and retired horses can be a fatal complication. For all horses, but especially

Water is indispensable, and affects all the horse's metabolic processes. Water helps to build new cells, and aids in digestion, excretion, thermoregulation, and immune responses. Insufficient fluid intake can cause a slow collapse of the entire internal environment, including thickening of the blood, rising blood pressure and heart rate, metabolic disorders, and colic.

older ones or those in poor health, it's extremely important to pay attention to fluid intake.

Because of this, providing buckets or containers for horses to drink from makes it easier for us to monitor their drinking habits—whether the expected amount is gone, or suddenly there is too much or too little remaining.

How do we catch a problem with fluid intake in time in a herd using an automatic waterer?

Systems for measuring consumption from a waterer do exist, but they're still fairly uncommon. Most people simply monitor their horses while cleaning the paddocks and pastures, observing how often the horses head for the waterer and how long they drink from it. Over time, it becomes easy to notice if one horse is appearing less often at the waterer, and the horse in question can be targeted for intervention prior to the onset of dehydration. Another time to watch for drinking patterns is after horses are fed concentrated feed. Most horses will have a drink of water after eating grain or concentrates.

The Other Extreme: Excessive Fluid Intake

If a period of extremely hot weather has started, increased water consumption is natural, especially if the horses are primarily eating dry grass or hay. There's no need to worry about a significant increase in water intake. However, if only one member of the herd is showing an increased need to drink, especially if the weather isn't particularly hot, we should take notice as it may signal an impending problem. For example, in combination with excessive coat growth, we should consider Cushing's syndrome—one of the most common disorders of the endocrine system in horses. Many senior horses suffer from this disease, in which the cerebral cortex overproduces pituitary hormones. Another possibility with increased fluid intake (and more frequent urination) is kidney failure, a state in which the kidneys are no longer capable of properly removing metabolic waste from the system. Another quite common reason for an unusually high intake of water is diarrhea. Due to excessive loss of water in manure, the body compensates with increased thirst, and with less frequent urination. Despite the horse's increased water intake, if it's a severe case, dehydration will eventually set in. Pay careful attention, daily and routinely, to the condition of horses' manure—not only in the paddocks and runs, but also in pastures, where loose manure can be easily overlooked in longer grass. Never underestimate diarrhea!

Tip: When Horses Reject a New Water Source

If your horses drink less willingly after transitioning to a new water source or reject it outright, measure the pH of the water. **Horses prefer water with a neutral pH (which means 7 or close to 7). Water with a high pH** (above 8) is alkaline and has a soapy taste, and animals will reject it. Water with a pH below 5 is acidic and will usually be perceived as sour and unpleasant, and your horses may drink less. If you're not certain of your water source or simply want to be careful, it's easy to measure water's pH with a piece of litmus paper.

If a horse's health condition requires a certain drinking regime, buckets can be used to monitor the exact amount of daily water intake, even if automatic waterers are usually relied upon.

Another option is to measure the water's hardness. A simple rule of thumb: the harder the water, the more alkaline it is. The softer it is, the more acidic it is.

In order for a horse's intestines to move food smoothly, it must be sufficiently moistened. Lack of fluids can lead to a gradual drying and hardening of stool, which begins to act like a stopper. Some of a horse's early symptoms of this kind of constipation are depression or reduced interest in his surroundings, and (although he'll continue to eat hay, making things worse) he will stop drinking. He will gradually stop eating as well, and will usually try different maneuvers like stretching, lying down, and rolling. This kind of impaction colic can occur gradually, and can go unnoticed for several days. It can occur not just in high heat but in winter as well, when some horses will only drink the bare minimum of icy water. This, in combination with a diet of only dry hay, can cause dehydration and constipation.

A classic paddle waterer—make sure it's relatively easy for the horse to depress the paddle, so you can be sure your horses won't drink less due to the amount of pressure required to obtain water.

Frequently Asked Question:
What about wild horses? No one warms the water for them—they break a hole in the ice to drink out of a cold stream. How is it that they can survive harsh conditions and travel huge distances to find water without colicking?

Horses should have constant access to water in their pastures and in paddocks. During the winter, if waterers or buckets aren't heated, hot water should be added once or twice per day to make icy water lukewarm. Research has been done by several independent teams confirming that horses given ice-cold

A home-made float waterer—a smaller container within a larger one, with the gap filled with insulating foam. Under the cover, there is a float that keeps the level of water constant.

Veterinarians indicate that when temperatures are around freezing, horses drink about three times per day. When temperatures are in the 80s, they drink 1–2 times per hour. If horses are on pasture on hot days, and are given water only in the morning and evening, they are prime candidates for dehydration and impaction colic.

water will drink up to 50 percent less water per day than those given water around 50 degrees fahrenheit.

Wild horse populations have a completely different lifestyle than our pampered domestic horses, even if our horses are being kept on pasture. They move continuously over much greater distances. Exercise is a horse's doctor and fitness trainer; it's the horse's modus operandi. And not beneath a rider, but slow, continuous movement, most often with his head to the ground from bite to bite. Also keep in mind that the wild horse population pays a very high price for its toughness in the number of deaths of weaker horses. Herd members that don't have the necessary iron constitution, who are prone to infection, injury, disease, or lameness, are simply food for predators—or they die of natural causes, through a harsh process of long-term pain, starvation, and dehydration. Essentially, you don't see wild horses who can't break through ice and drink frigid water because any that aren't capable

These lucky landowners have an underground spring—they had no trouble building their horses a pond.

of surviving under such conditions are already dead. We would never allow our domestic horses to meet such a fate... and that's not a bad thing.

But the struggles wild horses face are exactly the intention of Mother Nature—uncompromising natural selection. If every animal handled the pitfalls of the wilderness without problems, in good health, herds would grow uncontrollably and carnivores would starve, unable to catch healthy, strong horses. Whenever we find ourselves saying, "In nature, no one would provide that... and no one would secure this... they'd have to do this on their own..." what we are overlooking is that these things are part of the unkind and uncaring centrifuge of natural selection, which stirs wild herds and pushes weak and less well-adapted horses to the very edge of life, where they don't last very long. This is the only way a healthy and resilient population endures.

We could certainly let Mother Nature take charge of our own horses—and the final result would be hardened horses who could survive virtually anything thrown at them. It would be wonderful to establish a new, hardy breed out of these horses; but we would have to come to terms with horrific losses and deaths along the way, too. Would we have the commitment—or the stomach—for such an approach? It would mean leaving animals to waste away, not helping the sick or injured, and simply watching as weaker horses die without help...

Tip: What to Do Without Electricity

If remote pastures don't have electricity, there is—in addition to the mechanical options mentioned above, where animals pump water themselves—one more option: the classic style of hand pump that uses the kinetic energy of flowing water, the so-called pitcher pump.

A pitcher pump works by creating a vacuum. The sudden increase in pressure when the valve is opened and closed allows water to be transported to a place up to 25 times higher than the pump's water source.

Such an approach is not only something most people do not actually want to do, but, according to modern frameworks of ethics in animal welfare, it would be absolutely unacceptable. So the outcome of a colic episode (whether due to a low flow of icy water, or in connection with any other issue related to care) isn't going to be a tougher but smaller herd of healthy horses, but rather an empty wallet and a vet who feels right at home with our horses.

Water Quality

Cool water limits the growth of harmful microorganisms—we should keep this in mind in the summer when temperatures climb.

Microbial growth is slower in 50-degree water than in water that's been heated all day by the summer sun. For this reason, water buckets should be refilled at least every two days, and shouldn't be left in the sun. Keep in mind also that darker containers will heat up more than lighter-colored ones. Out in pastures, the most commonly seen water container is 250 gallons, made of white plastic. These can be placed in the shade, or otherwise protected from direct sunlight. There are also large tanks with attached automatic waterers available on the market—these should be kept in the shade and cleaned regularly to prevent algae from taking hold (if you use chemical cleaners once in a while, be sure to rinse thoroughly!).

Personally, I don't like the idea of leaving any kind of machinery where horses are unsupervised and at liberty. I would place any kind of waterer behind a fence so the horses only have access to the bowl and wouldn't be able to wander around the machinery. Of course, there are owners who have never had a bad experience with machinery left in pastures, agricultural and otherwise—but I'm inclined to avoid unnecessary risks, and would never place a water cistern in an area where a horse could injure himself.

Maybe this will come in handy: if you have insurance against injuries to your horses (each company sets these policies a little differently),

Excellent refreshment on hot days—an open water source right in the paddock. Horses are able to both drink and bathe.

Keep in mind that we don't always get permission to use natural water sources for a water supply or even for open access for our horses. The Environmental Protection Agency and local water authorities may be involved.

find out when finalizing a contract whether an injury caused by an automatic waterer would be covered, or would be considered a failure on the part of the insured to provide a safe space for horses, and therefore not covered.

Because horses prefer fresh water to stagnant, one good option (in addition to creating a well as was mentioned earlier) may be to use available streams on your land.

By law in my area, the only way to use this water without consent from the water management authority is to provide it in a so-called natural way—that is, without equipment or machinery, like pumps.

Be aware of the law! In some places, when it comes to surface water collection, the following applies: If a private citizen collects water from an open source (pond, lake, river, stream, flooded sandpit, pool...) to give to pets and puts the water in containers without the use of any technical equipment, she doesn't need any special permission. If, however, the water is taken using a pump, a permit is needed, no matter how much water is taken. The rules may be different where you are!

At the same time, horses must not damage banks, water management facilities, fish farming facilities, or alter runoff conditions, impair water quality, or harm the general interest and rights of others. So I can't make access to flowing water available to my horses indefinitely, and I need to be careful not to allow my horses to destroy banks or let manure contaminate the water. What may be allowed in one state or county isn't always allowed in others...

In general, unless a water source is on protected land, it's within the law to give horses access to a specific portion of a stream—for example, by creating an access point (with pebbles, gravel, or non-slip concrete pavers) so the animals don't trample the banks.

With horses, we must be wary of two kinds of pollution—first, the excessive load on the aquatic environment of nutrients from manure and urine, and secondly, the pollution caused by constant disruption of sediment. Damage to aquatic ecosystems can be eliminated by creating a vegetation belt.

Vegetation Belts

A vegetation belt is the area on the shores of a body of water abundantly planted with vegetation, between pasture and the water's banks, and should be fenced off from your grazing animals. This intermediate zone should be allowed to grow so that it can fulfill its intended purpose—so horses shouldn't be able to eat the vegetation or trample it.

This strip, if properly established, will serve to filter pollution due to storm runoff. The roots of these plants will also help to filter groundwater running from the pasture to the stream. In addition, this vegetation will stabilize the banks against erosion, and can protect the riverbed from becoming clogged with materials like shavings or hay blown by the wind. It will also provide a habitat for many animals.

Vegetation Belt Composition

This should always be based on what the site can sustain, and from the composition of species already native to the habitat. This may or may not include hornbeams, ash trees, maples (only non-toxic species! So no to sycamore, maple, red maple, and ash maple), elms, alders, willows (non-toxic varieties are an absolutely wonderful choice), and poplars, depending on your location. For the understory, you can select shrubs such as dogwoods, hawthorns, and willow bushes; check what may be native to your area and is horse-safe. Reeds and cattails can be used for grasses; excessive growth into pastures should be prevented with regular mowing. Don't forget herbaceous vegetation, either.

In order for the vegetation belt to serve its purpose, pay close attention during the first couple of years, cutting back and thinning vegetation as needed.

It's important to exercise caution if you're thinking about disinfecting drinking water for your horses. Chemicals used to target microbial sediment, biofilms and algae in water buckets or water distribution systems can also kill a horse's gut flora.

A pasture tank. A horse recently joined our herd who had had a bad experience with one of these: during a skirmish with another horse, he sustained an injury from his shoulder to his leg and had to spend several weeks at the veterinarian's clinic. This cost his owner not only a lot of stress, but also a large chunk of her savings.

A carefully made waterer, but I wouldn't leave it out in the pasture. It should be covered so the horses can only reach the bowl.

The area set aside as an entry point for your horses should be reinforced carefully in order to avoid muddy banks.

After these adjustments, any stream running through your pasture may, in the eyes of the law, be considered a natural watering hole.

Drinking Water Safety

Your horses' drinking water must not contain any kind of microorganisms, parasites, or substances of any kind that could endanger their health. The measure for this is the same as the parameters used for determining the safety of our own drinking water. Many surface streams are currently contaminated beyond safe levels. The conscientious owner should get any water source tested by an accredited laboratory before allowing horses access to it.

Microbial Pollution

Microbial pollution of surface water is determined by the evaluation of the level of fecal pollution and the presence of intestinal pathogens in the water source. A reliable indicator of fecal pollution which is watched for is *E. coli*. The source of contamination is usually a leak from a septic tank or municipal sewage,

The protection of water resources in the US is determined by the EPA; specific regulations may vary by state, and are intended to protect the quality and safety of groundwater and surface water resources used for drinking water.

Try to avoid creating heavy shade by the banks of the water. The more shade over the water, the less the water will be able to stay filtered and clean as it flows.

and agricultural runoff from manure piles or fertilization. Coliform bacteria can even form in a well if there are decomposing leaves or other biological material in it. Because of this, a well-sealing lid with a diameter that slightly exceeds that of the upper well ring is an absolute must.

Unfortunately, the days when it was possible to let grazing herds drink from ponds and streams without fear are long gone. I repeat: by law, it's usually the landowner's duty to provide safe drinking water for the health of horses.

Determining the nitrate content in your water source is also very important. Nitrates on their own aren't toxic to horses, but the enzymes in the horse's digestive tract interact with them and make them quite toxic. Horses can be quite sensitive to poisoning by nitrates.

Providing water with high nitrate levels can have a number of negative effects on a horse's health, from problems with digestion and poor absorption of nutrients to direct poisoning (acute or chronic).

At high concentrations (for example, after storms cause runoff from fertilized fields), there is a real risk of a condition called *methemoglobinemia*. This is a condition where the nitrates absorbed into the blood oxidize hemoglobin to methemoglobin; iron in the blood then loses its ability to transport oxygen.

Acute poisoning manifests in its effect on the central nervous system and blood vessels (convulsions, paralysis), increased heart rate, and decreased blood pressure. In advanced stages of poisoning, the horse won't be able to stand, and death follows soon after. If your horse has access to surface water, be on the lookout for signs of elevated nitrates in the water after heavy rains (due to agricultural runoff), including diarrhea or excess fluid in otherwise normally formed stool.

Although adult horses are sometimes able to recover on their own from exposure to nitrates,

A good rule of thumb for planting around horses: we should use plants that are not going to cause any issues for horses. In addition to toxicity, we should also take into account the plant's overall "difficulty"—for example, thistles in a horse's mane are an unwelcome hindrance...

Use of streams for a water supply are within the limits of the law in many places—if the water flowing to the horses then flows freely back to the stream. This way, horses have access to water that's always cool, fresh, and clean. They also won't enter the stream and therefore won't damage river banks.

foals are far more susceptible to serious injury or death from high nitrate levels in water.

Keeping Water Containers Clean

Water containers should be cleaned every other day during the summer, and automatic waterers should be cleaned and also checked daily (morning and evening) to be sure they're working properly. Be sure to check for drowned mice or birds during dry weather!

Hay, grass, and forage residue are breeding grounds for pathogens and fungi. Water containers need to be scrubbed by hand to remove biofilm bacteria (algae and slime), as disinfectant is more or less ineffective against this kind of bacteria, in addition to being harmful to a horse's microbiome. And check containers regularly to be sure there are no sharp edges where horses can cut themselves...

Disinfecting with a UV Light

UV lights don't add anything to water or change its composition. They work by killing coliform bacteria and other microorganisms with ultraviolet radiation as the water flows past them. One disadvantage, however, is that any bacteria that settles in the pipe behind the lamp will be untouched. I know of one case where a UV lamp was the needed solution for foals who were suffering from prolonged cases of diarrhea—the culprit was the lamblia protozoan, which was killed by the UV light.

Location of Water Stations

The spot you pick for watering depends on how water will be added—if you'll have to do it manually, the best option is to save on labor and consider how far is far enough to travel several times per day. If water will be refilled automatically, consider savings in terms of dollars—the cost of digging below the frost line and laying pipes increases significantly as feet are added. Plan the best site for your water station in the paddock well in advance; it'll save you from countless headaches if future rebuilds are needed.

If you also want to heat the waterer in the winter (which I highly recommend), you'll need to also take into account convenient access to electricity. For the above ground part, high-quality insulation should be added to all connections. For unheated water containers that only function at temperatures above freezing, we can always add water from an above-ground hose—no need for digging. But keep in mind that although this method might be easier to install, the water temperature won't be as stable. When the hose is heated by the sun, the water will run far too hot for safe consumption at first. If there's some distance between the faucet and the container which needs filling, you'll go through a lot of water before it runs cold. I'd urge you not to just let the water run, but rather use it to water surrounding vegetation and trees, or catch it in buckets to use somewhere else.

Whether the supply line is above or below ground, always consider how the station will be oriented, especially if you have a large number of horses.

Horses will often travel from the pastures to the paddocks in large groups in order to drink. If there's a single water trough for a larger herd, there can literally be a stampede around it. In order to prevent potential disagreements and injuries, place the waterer so that none of the horses are pushed into a corner. The ideal location is either centered on a long wall, or in a completely open space that can be approached from all sides.

Another factor to consider is the presence of trees. On the one hand, trees can provide cooling shade, but on the other, having a tree overhead will mean leaves and bird droppings—which in turn means frequent cleanings. A sheltered position for a waterer is always a good idea, not only because there will be less dirt, leaves, and so on blown into it, but also because the water won't be further cooled by a blowing wind (especially in containers with a large open surface area), which can save on the energy needed for heating.

The height of a container also has a significant impact on how dirty the water becomes and how much gets spilled. If the container is low to the ground, many horses will be tempted to paw the

Be on the lookout for contamination of the waterer by manure. Check waterers daily!

water, getting it dirty or even tipping the container over. Therefore, a raised site can be an advantage. Another genius solution is placing the container behind a fence—see the photo on the right.

Also, it's worth it to make a small investment in building a grip on the edge of the water tub, as hoses left unattended easily slip. Keeping your water pointed in the right direction will help you avoid a lot of frustration.

With automatic waterers with a bowl (whether float, paddle or pin), there's no danger of a horse pawing and kicking them, but they do offer an attractive surface to scratch against. In order to avoid unnecessary injuries and to protect the bowl itself from damage, I recommend installing a protective ring around the bowl. This works equally well for wall-mounted and floor-mounted models.

A paneled water trough isn't just an aesthetic choice, it's also safer. Horses can't be injured by metal corners. I recommend wrapping sharp corners with slightly rounded timbers.

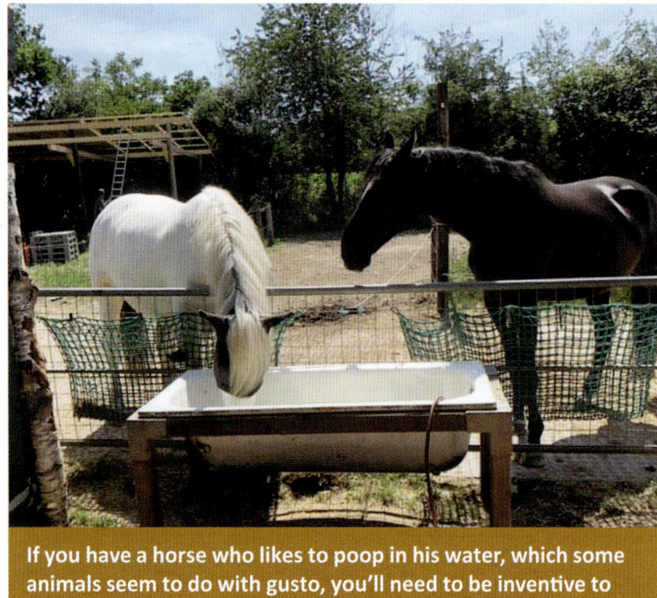

If you have a horse who likes to poop in his water, which some animals seem to do with gusto, you'll need to be inventive to outsmart the bad actors.

Water, Not Ice

Winter is dreaded by so many stable owners, often due to the excessive effort needed to make sure horses' water doesn't freeze over.

If I still haven't managed to convince you to buy a heated water dispenser or container, and you plan to try to maintain a regular tub of water in an icy pasture, I have a few tips to help postpone the inevitable freeze:

- Protect your hard work by keeping manure out of the water. You can surround your waterer with a stone or wood barrier that allows a head and neck to reach the water, but not a horse's rear...
- Place branches, balls, or empty plastic bottles in the water. When the horses poke these objects, that will break up any ice beginning

In the summer, it's easy enough to keep a water tub filled, but once it's cold out, hoses can't be used reliably. You could carry water out in barrels or other containers in a wheelbarrow, or you could make life easier with a heated bucket.

to form. If it's windy, a ball being blown around will keep the water moving so it'll take longer to freeze. However, in severe cold, nothing will help except heat. Also, if you use bottles and have young horses, they may, while being playful, grab the bottles and remove them from the water, which defeats the intended purpose...

- Place bottles of hot water in the containers—this will slow down cooling. Be careful to use plastics that can withstand high temperatures to avoid chemical leaching.
- Place an electric heated pad under the water container (if done safely). These can be found at farm supply stores. If your container is too tall, the water at the top will still freeze, even if the water below remains liquid.
- If you use tin or cast-iron tubs, you can place candles underneath to warm the water slightly. If you're going to try this, you'll need to make a lining or screen so wind won't extinguish the candle, but it needs to allow enough air to keep the flame going. (Never use candles inside a structure!)

Keep candles out of reach of the horses. Here, they're on the outside of the fence.

- An old boiler can be used for water—frozen water will warm up very easily with a heat source placed underneath.
- Some owners use aquarium heaters in the water tub (in a PET bottle, PVC pipe, or something else to protect it from damage by horses).

For anyone tempted to create a DIY electric heater (and also for any open flame), remember that horses are curious creatures and will "explore" anything new with enthusiasm—be safe! The complications from a candle's flame in an outdoor setting in winter are probably minor compared to the possibility of electrocution. Any flaw in the design of a home-made gadget could lead to a deadly situation for your horse!

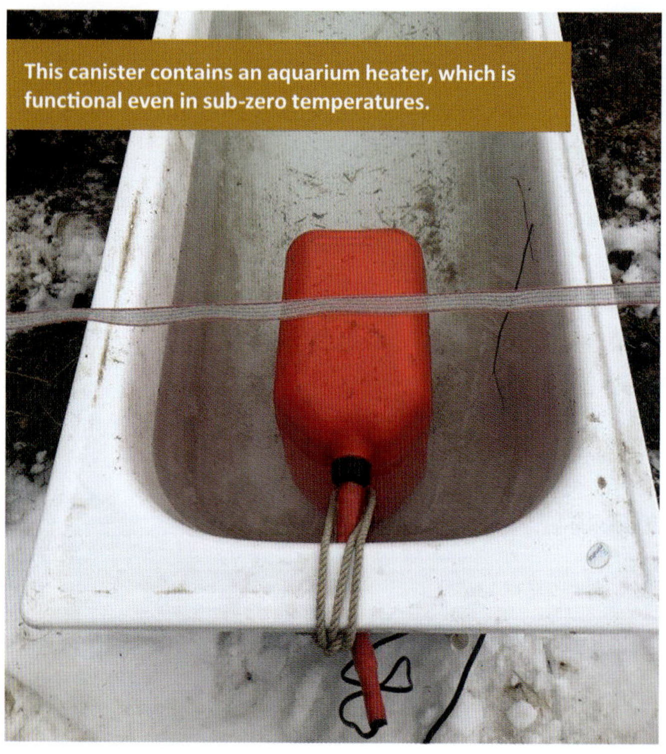

This canister contains an aquarium heater, which is functional even in sub-zero temperatures.

A final piece of advice: don't rely on snow as a substitute for water. Depending on density, a horse would have to ingest 5–7 times as much snow to achieve the same hydration he would get from the amount of water he would normally drink. Basically, he would need 7 buckets of snow for each individual bucket of water. Does this make it clear why it's irresponsible to expect horses to use snow so you don't have to provide water?

Heated waterers are a perfect solution for outdoor horse setups. This model with a pedestal costs around $450 dollars. Of course, more money will be needed for excavation work, pipes, and so on, but once setup is complete, ongoing costs should be low, and multiple problems will be solved.

chapter VI.

"URINATION STATIONS"

Sometimes a beautiful paddock is built, and yet the most important mud-fighting element is forgotten. In fact, something essential to keeping mud at bay is often missing in these carefully prepared areas. Yes, that's right: your horses need a place where they can comfortably pee.

Unlike passing manure, for many horses, urination is something they need to do in isolation—they'll often stay away from other horses when they're going to pee. A horse can pass manure while walking, trotting, or cantering, but can only urinate while standing still. In addition, when horses are emptying their bladders (which, under normal circumstances, happens several times a day), they need to consciously relax their muscles. Even a stressed horse will poop if he needs to, and sometimes more than usual. The same horse, however, will need to wait until his nervousness subsides before he'll be able to pee.

Can We Make It Easier for Our Horses to "Go"?

Of course! Although some horses will go just about anywhere, most prefer a soft, absorbent surface which keeps pee from splashing everywhere (namely, on their own legs and bellies).

Most horses have opinions and preferences about where they like to go—and most of these horses will prefer fluffy stuff to packed and trampled surfaces. In a hard paddock, peeing in a big fluffy pile of hay will seem like the best idea ever! Of course, their owners would rather they didn't, and will do everything possible to save the hay from this undignified fate. But all this work to save hay and money can seem to favor our wallets over our horses.

After a passing shower, these horses came out of their shelter to relieve themselves—the chestnut on the sand, and the gray on the straw. Horses like both surfaces for this job.

It's not a huge stretch to say that being careful with hay can lead to carelessness toward the horse. Unfortunately, some outdoor stables have shelters that aren't set up for the horses to lie down, so they don't have bedding, but they don't have any grass or mud either. With nothing but hard surfaces available, the horses are forced to pee on unsuitable surfaces. You can help them easily by setting up a spot for them to go—let's call it a "urination station"!

Is There Any Difference Between This "Station" and a Bed?

Honestly, no, except for size. Horses will always go in straw if it's available; they don't care what we call it. Since a pile of absorbent material will always attract them, let's use that to our advantage! Anyone who spreads straw in a paddock will tell you how much horses enjoy it—they rummage through it looking for treats to eat, and they lie down in it, roll in it, poop in it... Horses don't differentiate between bedroom and bathroom.

Folks who love to spoil horses don't see a problem with this, and they'll happily go through straw as quickly as they would if they were keeping their horses in a stall, simply taking it all away and replacing it as often as needed.

Toilet and Bed in One?

Some people like to lay deep bedding for horses in winter and take it away when the weather warms up again; while it's there, it provides a place for the horses to lie down and pee—bed and toilet, all in one.

However, maintaining deep bedding isn't really feasible for more than a few horses unless you have the help of a tractor, especially if there's no covered area. And what about the many people who only have the use of their own hands, a wheelbarrow, and a pitchfork? Maintaining any kind of bedding would require daily cleaning, and the area would have to be small enough that it wouldn't really be worth the work.

Rather than use a stingy amount of bedding, which the horses will also use as a bathroom anyway, it saves you both money and time to set up a spot large enough for a pee pad, but too small to lie down in, making it easy to maintain.

Frequently Asked Question:

I have hay bales for my horses, and I spread hay for them as well. How do I keep them from peeing on it? I don't want to waste good forage, but at the same time, I want them to be comfortable. (Our stable doesn't have flat surfaces outside.)

Not having flat surfaces doesn't mean you can't have soft surfaces (like grass or even clay), which may make perfectly good spots to pee. Most paddocks still have a few comfortable options even without bedding, whether it's clay, sand, or fine gravel. Of course, if these areas get trampled down and turn hard (which can happen, especially in the summer), then there's nowhere that's serviceably soft.

In this case, horses will use any soft material they find, even if it's their feed...

Try to re-train them to another, more suitable (for you) area. Provide them with a small,

Although urination on a hard surface is fairly uncommon—usually the last resort of older horses who aren't as mobile, or horses who were spoiled by a permissive mother or lead gelding—you shouldn't allow horses to become accustomed to peeing only on solid surfaces. Make it a priority to give them a place with absorbent material.

Toilet and bed in one—used extensively for both purposes.

If your basic paddock is sand, there's no need to build an outhouse. In this space, they can comfortably go anywhere—sand is absorbent, and therefore a popular option.

easy-to-maintain, absorbent space that will be attractive to them: create a defined area and fill it with some kind of soft material: straw, sand, leaves, sawdust, or shavings. You'll be happy knowing you've helped satisfy one of your horses' most basic needs.

If your basic paddock is sandy, there's no need to worry about creating a dedicated potty spot—sand is a favorite and they'll happily use it.

Building the "Urination Station"

On Natural Surfaces

If you want to build your outhouse on a clay or grass paddock, there's no need to worry about a base—the ground will be absorbent enough. You only need to decide whether you'll just "throw down" your absorbent material, or want to separate the area somehow. Going with the "throw" approach will work best in a larger field or pasture; where there's enough space, it won't matter if your potty material gets scattered by the horses or blown around. In a smaller space where it'll be necessary to clean up the mess each day, it's probably best to create a defined perimeter around your station. Wood works perfectly for this. I've seen concrete curbs used, but I've also seen a horse badly hurt himself by rolling too close to one of these.

- When using logs, think about how to securely anchor them. I suggest finding ones with a large enough diameter that even after being put in a trench deep enough to anchor them, they'll still stick out enough to keep your "toilet paper" from blowing everywhere. This also encourages horses to lift their legs as they leave so they drag less material out with them. No one wants toilet paper stuck to their shoe when they leave the bathroom! If you're using logs with a smaller diameter, don't dig them a deep groove, since they won't protrude very far. Instead, use wooden pegs to anchor them.
- Don't even think of using metal hooks as anchors—when they eventually loosen, they can cause serious injury!
- If you're willing and able to spend the time and money—burying logs standing up is an option that's durable and looks great!

Every bed will at least occasionally be used as a potty.

On Paved Surfaces

On a surface with grass pavers, you'll have several options for anchoring the barriers of your potty area. Thanks to the fact that these pavers have holes for grass to grow through, it's easy to anchor something to the substrate. Whether you use ground screws, homemade anchors, or simply make a fence that, due to its size and weight, won't need additional anchoring, always consider the horse's safety first.

What about concrete surfaces? These can't be truly anchored without using a special drill, and there will be no absorbency *under* your urination station no matter what you use to cover the concrete, so lots of urine will have nowhere to go but everywhere.

It will drain depending on the slope, creating messy streams of pee.

Takeaways:

- Best case scenario: be mindful of this ahead of time, and leave your chosen area unpaved.
- If your basic paddock is already completely paved, you can carefully install an impermeable pond foil. You'll need to take care to keep it deeply bedded to protect the foil from damage.
- Use a barrier that's heavy enough to stay put even without anchoring. You may also be able to skip the impermeable layer, if you bed deeply and use absorbent material. (Keep in mind this will mean you'll need higher barriers.) In this situation, straw definitely won't work, as urine flows easily through it. Your bottom layer will need to be very absorbent: think sawdust, fine sand, granular bedding, or clay. These can be used together with manure for compost and subsequent fertilization of pastures. You can add straw as a top layer for aesthetics, and also to stabilize the bottom layer, if you want.

If it's a larger space that's used as the potty, remember that your horses will probably lie down here as well as use it for its intended purpose. I personally don't mind this, but it does mean it needs to be kept clean. I bed with straw, so I simply clean it out every day, just like a litter box. If you use sand, it'll stay mostly maintenance-free for a long time before it starts to smell, which will be your signal that it's time for a change or a top-off. This kind of potty can be essentially indistinguishable from a nice bed.

I encourage you to look at your basic paddock through the eyes of a horse and ask yourself these questions:

- Is there at least one area with soft, loose footing like clay, sawdust, or sand, or is it all paved?
- If my horse needs to pee, is he going to be able to do it without it splashing back on him?
- Is this place accessible? Is it easy to get to and into?

If you can answer yes to all these questions, congratulations! Your horses have a comfortable basic paddock with a potty!

chapter VII.

OUTDOOR FEEDERS

Although there are plenty of advantages to outdoor stabling, there's one problem that you'll need to address: feeding hay. When horses are fed in stalls and only go outside for a few hours, no one really worries about the hay in the paddocks. A stalled horse will eat forage throughout the night and will have a snack (also indoors) in the morning. During the day, they may amuse themselves picking at whatever's available outside, or they may get a little snack flake. If it's a wet or windy day, most hay fed outside will be given up for lost. This isn't a big deal when there's a dry pile of hay waiting inside, but for our horses who are being kept outdoors, it's a completely different story...

For our outdoor horses, there's no "pile of hay waiting," but having our horses outside is the ideal. For a horse that has to withstand winter outdoor temperatures both day and night, non-stop access to hay is imperative. It's also far more important to manage unnecessary hay waste. But how?

Frequently Asked Questions:

- *Can we be sure that our horses have constant access to roughage on cold winter days or during bad weather, while avoiding waste and ensuring that the horses don't end up using hay for bedding?*
- *How can we best monitor the quality of the hay when it can become moldy before it's touched by the horses during rainy stretches?*
- *How can we prevent our more crafty horses from burrowing into a hay bale, where they can breathe in far too much dust and develop respiratory issues?*

Burrowing in.

For as long as people have been keeping horses outdoors during muddy seasons and through winters, they've been looking for a good way to feed them hay. The need for something that allows horses constant access to forage, prevents them from making a bed out of it, and limits unwanted burrowing and over-eating prompted the creation of feeders.

Fence feeders can be used with or without nets.

There's no such thing as a maintenance-free feeder!

Absolutely every type of feeder needs some kind of attention. Whether you make a simple box, or you buy a well-constructed feeder with a roof, you'll still need to be sure to clean out leftover hay from the bottom and other hard-to-reach places so you don't wind up with rot or mold. Feeders without a bottom can simply be moved to a new place, but you'll still need to pick up a rake and clean the old spot...

Hay nets can be fully emptied by a horse, so no need to sweep up debris, but they're also susceptible to tearing. Horses can create holes in nets with surprising speed—they'll need to be checked and repaired often.

Ring Feeders

In the past couple of years, ring feeders have become almost as ubiquitous as hay nets. They're found in a wide variety of designs:

- Metal rings with gaps
- Solid metal rings
- Concrete rings
- Homemade wooden rings
- Commercially made plastic feeders

When shopping for a feeder, pay close attention to the design. Don't be tempted to buy a cattle feeder for your horses. The design doesn't take a horse's safety into account. Also remember that a horse, unlike a cow, won't stand calmly if he gets a foot caught in the feeder...

What to Look Out For with Ring Feeders

- If it's metal, pay close attention to the design—make sure there are no sharp edges or protruding screws, and make sure all the welded seams are smooth in order to avoid cuts or lacerations.

- With ring feeders, there's always a chance that a horse could get a leg caught, and horses can easily panic.
- Make sure the shape and height of the lower edges are appropriate for horses—again, this can present the risk that a horse could injure his front legs.
- For the type with gaps or mesh, keep in mind that there will be a bit more mess and waste. Unlike feeders with solid sides, hay will fall out.

Square Metal Feeders

These feeders are already available on the market in several variants. Some of these are more suitable for horses than others, but as with ring feeders, sometimes sellers will offer a feeder as suitable for horses when it should only be used for cattle.

Of course, it's not necessarily ideal to use a large metal structure around horses, but as with anything else, there will be strong opponents and also those who've had no problems after years of using them. Some have even built their own square or rectangular feeders, on their own, quite successfully.

Hay Boxes

There are "homemade" brands, and these can also be made to order by carpenters. I've seen some made from plastic fruit and vegetable crates. There are also safe and durable plastic crates that are commercially available.

With any feeder that has gaps, expect minor losses.

This horse became trapped in a ring feeder, and the owners were unable to free him until he finally fell (after trying to run while entangled). His injury was treated successfully, however, and he healed.

This homemade feeder has vertical bars spaced far enough apart that there's no risk of a horse getting a hoof stuck, but close enough together that the horses can't burrow into the hay.

Feeders that are too low can tempt horses to climb right inside.

A wooden vegetable crate which has been modified to be used for extra forage in pastures—it's easily moved from place to place so the ground doesn't get unnecessarily trampled.

This crate has been reinforced with metal edges to protect against chewing and has a net to conserve hay.

A repurposed fruit crate.

A feeder can be quickly made from pallets and a hay net.

Commercially available covered feeder.

Bell Feeders

Bell feeders can fit over a large bale. Horses can rub their manes on the opening of this type of feeder, but not everyone who uses them has this problem, and lots of folks are happy with them. These feeders are a durable, high-quality option, which can mean a higher price tag.

Corner Feeders

In Shelters

Corner feeders are a simple to use option outdoors (or in a stall)—they require less work and don't require us to put together some ornate structure. They're easy to install, requiring minimal strength or skill with tools, and yet they're underutilized. These save a lot of hay—horses who use this kind of feeder don't tend to fling the hay around. They can't climb into it, and even without a net, they don't usually toss much on the ground.

Outdoors

If a pasture or field has no shelter (because it's only used for a short time during the year), outdoor corner feeders can be created. In the basic paddocks, these feeders should be built against a fence or wall and as far as possible from other feed stations and shelters in order to encourage horses to use the entire paddock.

A corner feeder in the paddock or pasture provides a dry, sturdy (since it's supported by a wall) shelter, so horses are encouraged to eat from the hay net inside.

Large Capacity Feeders

These homemade products aren't just on the ground locally—you can travel anywhere in the world and see giant feeders where it's possible to throw many bales at once. They're usually made mostly out of wood, and if you stop to chat with the owners about the pros and cons, you can avoid rookie mistakes, as well as pick up some interesting ideas.

Covered Feeders

You can incorporate a large feeder as part of your shelter, to keep your horses and your hay out of the rain.

Sliding Feeder Gates

Sliding gates can be adjusted (and then anchored to prevent sliding) to the dimensions of your horses—so that the gaps between the bars are wide enough for the horses' necks, but not their shoulders. They are made from many materials, but often can be purchased from metal manufacturers.

Horseshoe Feeder

(Shown in the photo above.) These feeders make it possible to load hay into a center area that's out of reach of the horses, so whatever mixture we choose can be put in front of them. Whatever they throw back in the center can be easily returned to them with a simple toss of a pitchfork, having been kept safe from manure or trampling.

Nets and Other Slow Feeders

All the feeders that I've discussed so far can be combined with netting, either small or large mesh, which does the job of preventing hay from being thrown around, dug into, or lost due to wind. Also, these nets can help slow down a greedy eater if the mesh is small enough. If your horses are thriving perhaps a bit too well, that might call for a slow feed net. Since horses are capable and tenacious animals, count on them to learn how to win against a cheaply made net.

You can use a search engine or find a horse discussion group to learn about every kind of hay net, including custom-made nets from small businesses and independent sellers. Here you'll also find reviews: for some companies, the references will be great; for others, not so much.

Before you invest a chunk of money into lots of hay nets, be sure to carefully research the manufacturer regarding the durability of their product.

Keep in mind who will be handling and loading the feeders. If you don't have the help of large machinery, look for products that are easier to handle (for example, the bell feeder). You'll be able to use most of these in conjunction with a hay net.

Tip: Acclimating Your Horses to Hay Nets

For the first couple weeks your horses are using hay nets, be sure to also provide loose hay so they aren't hungry (and therefore stressed) while they learn how to pull hay from the nets. It's important in the first few days that the horses are full already while using them, so they can see it as a fun distraction and can work at the hay nets calmly. In addition to making your horses happier, this will also help to prevent damage to your nets due to "hangry" horses.

Mesh Size

Anyone who works around horses regularly will tell you that horses learn very quickly how to pull hay through 1.5-inch mesh nets, and some find it so entertaining that they'll choose to eat from the hay net even when loose hay is available. Even foals can be interested in hay nets, but be careful! With the little ones, it's especially important for the hole size to be small enough that they can't get their hooves stuck. An adult horse might be safe with 4-inch mesh, but a foal won't be. A leg caught in a hay net is an avoidable, but common, "accident" that can result in serious injury.

Trying to save money or time by using a hay net alone as a ground feeder not only leads to quicker destruction of the hay net and more wasted hay (from trampling, scattering, and peeing), but also higher risk of injury from a stuck shoe. This kind of attempt at saving will only cost more in the long run—always have a fence around the net at the very least. The best option is to combine the hay net with a feeder.

Most horse owners who use some kind of device to slow their horses' hay intake, from nets to metal grates, will hear two objections:

- It's unethical.
- It has a negative effect on the horse's teeth (with metal grates).

Ethics

Some feel it's unethical to limit a horse's food intake. They'll ask whether it isn't similar to what's done on feedlots, where horses desperately hunt in their feeders. The answer is another question: Is it better to let a horse become obese? Although an obese horse clearly won't suffer from hunger, his condition poses a great risk to his health.

For our horses' well-being, sometimes slowing down their need for speed is indeed ethical—and unfortunately, quite often, it's simply necessary. Not everyone can dole out hay whenever needed throughout the day. Even if your animals are at home, other commitments like work may make it hard to get to the horses more than once per day. If any of the horses tend to speed-eat, are insulin-resistant, or just struggle with their waistline, the only feasible option is to find a slow feed option that will work for them, like putting a net over an entire bale (which will ensure that the hay lasts, but horses will move around less), or hanging a few flakes here and there in smaller nets (to encourage horses to move from place to place).

Slow Feeders and Teeth

Equine dentists were caught off guard by the sudden increase in the number of people using metal grates as slow feeders on feeding crates.

This grid, although metal, hasn't caused any problems for horses' teeth, according to their owners after regular dental check-ups.

Research out of Europe shows that when you use this kind of grate, the horses' teeth come into regular contact with the metal while chewing, causing excessive and irregular wear.

If they have room to do so, horses will primarily use their lips to take mouthfuls of hay. If the feeder doesn't have enough space, they'll start to use their teeth to grasp at pieces of hay. If the spacing is inappropriate for your horses, the entire surface of the front teeth will come into contact with the metal, causing real damage to tooth enamel. Although it's not the rule, dentists have reported these problems, so don't underestimate the risks posed by using metal grates and be sure to schedule for frequent dental checkups.

Because of this, plenty of people have chosen grids made out of other materials—like plastic—instead of metal. These can be found sold individually as well as part of a set with a feed box.

Using smaller hay nets is labor intensive, especially in terms of the time needed to fill them and hang them in various places around paddocks and pastures. Although some people will never give up these nets despite these difficulties, plenty have turned to easier-to-fill options like feed bags or barrels.

Homemade feeders made out of various materials are only limited by your imagination, but always pay attention to how the horse will hold his neck while eating.

He shouldn't have to lift his neck—which allows dust to get in the horse's eyes and causes his back to drop uncomfortably—or twist it into unnatural lateral positions, which can lead to over-muscling and blockages over time.

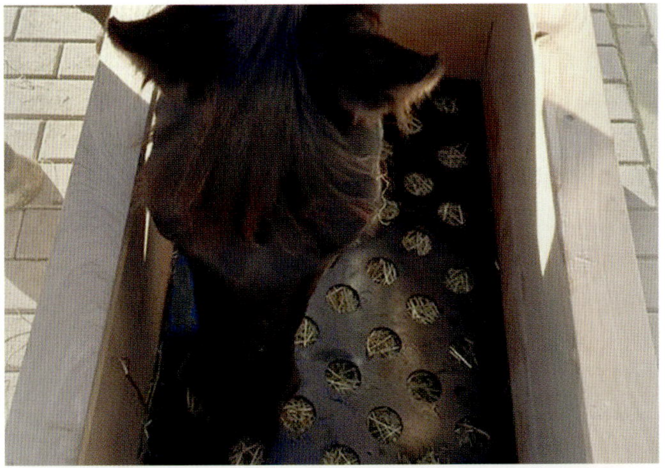

If your pony shares his feeder with larger horses and struggles to reach the hay, there's an easy solution.

So what works? The ideal type of slow feeder allows the horse to eat with his neck completely straight and his head lowered.

Learning by Example and Trying Other Gadgets

What tricks can we learn from other people's experience with homemade feeder design? For starters, if we're going to build a permanent feeder, its diameter should be about 10–12 inches wider than the widest bale, so that a hard-working horse can't pull the hay over the edges of the feeder.

Lots of people like a feeder to be big enough that hay falling from a horse's mouth will fall back into the feeder, not behind it. Also, **be careful with the height of the edges of the feeder—if the edges are too low, it's easy for horses to climb in.** Not only will this allow them to ruin their dinner with urine and manure, but they run the risk of hurting themselves. An edge that's too high, on the other hand, will make it hard for them to eat hay at the bottom.

When building, think about how you'll protect the hay from rain. Add a roof or canopy that's large enough and has sufficient overhang so the horses can fit under it without the water running off onto their backs (which, unfortunately, I see quite often). If it has to be a smaller roof, place the feeder at the edge of the paddock, with the roof sloping away from the horses and towards the fence, so that runoff won't splash the horses' backs.

Of course, the best solution for rainy days is a feeder that's covered so well that it provides shelter for the animals too. Whatever size roof you decide to build, if you're in a cold climate, be sure the bearing

capacity can manage the weight of whatever snowfall might come your way.

The more spacious the shelter, the happier the horses...

The difference between a covered feeder and a feeder placed under a shelter's roof can sometimes be eliminated—**if there's enough room for both the hay and the horses surrounding it to stay dry.** Of course, not every feeder is built so that it can be approached from any side: some, in fact, need to be set up so horses are prevented from standing in a certain place (by a neighbor's fence, for example).

Whether you decide to build a feeding station which can be approached from any side or one that is only accessible in the front, always make sure that there are no protruding screws, nails, or knots of wood. Sheet metal corners shouldn't have sharp edges where a horse can lacerate himself. Check that there are no gaps between planks large enough for a horse (or foal) to catch a hoof—this could result in a broken leg if a horse rolls or paws by the feeder. **Finally, take care with halters and blankets that clips are positioned so they can't snag on the nets.**

Tires

Feeding hay out of tires may seem a bit out there, yet plenty of farmers use them without any problems—the bigger the tire, the better. According to available studies, the material (from used tires) is harmless, but if you don't trust this, be sure your horses are "disciplined" eaters—no biting or licking of their feeder. If it turns out you have rodents getting into the tires, you can always remove them again! Plenty of people report being happy with their tire feeders even after several years.

Hay nets can be easily added to prevent horses from pulling the hay out of the tires and to slow

A hay carousel not only extends eating time thanks to the mobility of the nets, it also encourages the horses to move as the hay nets slowly turn. This idea originated in Germany.

intake. Another huge bonus is the low cost. They're very popular in Germany, where you'll find them in plenty of paddocks.

Hay Balls

A hay ball can be left loose on the ground, but because they don't hold much fodder, they shouldn't be relied on for more than a bit of entertainment. These can be useful left out in pastures for horses that grow bored easily—some horses enjoy mining the hay out of the small holes.

In the category which includes things like hay carousels, we should discuss baskets and containers which aren't primarily intended for horses, but which make appearances in paddocks and pastures. For example, I often see large water tanks (approximately 250 gallons) in metal frames being used as feeders. These can serve the purpose well, but beware sharp edges and joints, and be sure that there are no spaces that can catch hooves! There are plenty of ways to modify bins and bags for use in a paddock, but it's also quite easy to find something suitable at the store.

People have been creative in their use of a variety of materials to build feeders for their horses—some seem quite brilliant, others strange. But at the end of the day, if it's safe and it works, that's all that matters.

chapter VIII.

SALT LICKS FOR OUTDOOR HORSES

Horses need salt, but some guidelines need to be followed. For example, salt should be kept out of reach of young foals if they're overly interested in it.

Providing salt licks in outdoor areas seems like a small thing, until you think about it. Imagine purchasing an expensive, organic salt block with several essential minerals—suddenly it seems careless to just put it anywhere, a helpless victim to the first heavy rain. But even if you don't give your horses anything fancier than plain salt, you'll want to put it where it can be used by the horses, not washed away.

The most obvious option is to hang our salt licks or mineral blocks under the roof of a shelter. Don't hang them directly on a wooden beam, since both humidity and the horse's saliva can cause the salt to soak into the wood, creating a delicious salty temptation where there was once just a plain piece of wood. Anyone will tell you that trying to protect wood from being chewed once it's begun can be a huge chore.

To the rescue—the salt lick holder. The invention and placement of the DIY holder can be a pretty popular pastime; there are plenty of creative and inventive options.

Is It Okay to Give Horses Unlimited Access to Salt? Absolutely. The only exception to this is small foals, who can develop diarrhea from too much salt intake. If you have a foal that is too interested in playing with the salt licks, make sure to put him out of reach for the first few weeks of his life. The licks can be returned to their usual space gradually—beginning with a couple hours and gradually increasing the time. The foal should eventually learn to listen to his self-regulating taste mechanisms, at which point you can stop limiting the availability of the salt lick. If at any point the foal suffers from a bout of diarrhea, take careful note of how much time he spent with the salt lick and consider again removing the lick for part of the day.

Sodium chloride (NaCl), or rock salt, is one of the most important substances for vital activity and smooth functioning of a horse's body. All mammals need at least some salt; it ensures blood viscosity and is used in muscle contraction, as well as regulating blood pressure and production of gastric acid. It also takes care of, among other things, the body's pH levels—keeping blood values in the (very narrow) optimum range of 7.36–7.44 on the "acidity-alkalinity" scale. Any increase or decrease in pH outside the optimum range is associated with changes in enzyme activity, protein disruption, oxygenation disorders, and other serious problems.

Salt is essential for a horse's body to function. It regulates the body's water management, transmission of nerve impulses, and muscle contractions. Its lack is immediately felt by a horse as a desire to seek out salt. Wild horses will wander miles to find a spot with a salt reserve. This natural need is filled by searching for saline soils and deposits of salty minerals. In the case of domesticated horses, we provide for this need by having salt licks readily available.

Salt occurs naturally in the form of mineral halite, or rock salt, which is formed by the crystallization of seawater in isolated bays and drying sea shoals.

Manufacturers offer all kinds of shapes, and even different flavors. As for the addition of flavors, for most horses this isn't something we want or need, as most horses will naturally seek out salt without any encouragement. For horses that we'd like to see visit the salt lick more often, we can consider providing them with a flavored option or a toy (like a jolly ball or activity rope) which incorporates a salt lick.

An adult horse needs approximately 3.5 grams of sodium and 2 grams of chloride per day. Sodium content in pastures is often deficient, so supplementation becomes an absolute must.

Horses can become deficient in sodium and chloride as a result of excessive sweating or salivation. Horses will seek out salt when they lose sodium, for example after a long trailer ride on a hot day, or after strenuous exercise like an endurance competition. Therefore, no farm should be without access to salt for the horses, no matter the size or discipline. If you see horses licking walls or other horses, or eating sand and clay, consider the possibility that the cause may be insufficient salt intake. Long-term deficiency is manifested by loss of appetite, muscle weakness with increased sweating, and reduced performance. The horse's coat may appear coarse and brittle. On the other hand, excessive salt intake is usually signaled by diarrhea. I can't emphasize enough that if you have newborn foals, be sure to place salt licks so that the youngest foals can't reach them.

It's important to know that too much salt can be toxic, and in extreme cases, can result in death. The recommended maintenance dose for horses is 5 grams of salt per 250 pounds of the horse's body weight.

Salt, although one of the cheapest supplements, is one of the most important. Be sure to place plenty of salt licks of differing designs in various places around their basic enclosure. This pays off especially with horses who may be less enthusiastic about their salt intake: there's a better chance they'll find their personal favorite type of salt lick.

After giving your horses salt, they must be able to drink! Make sure that where the horses have access to salt, they also have easy access to water, as salt will make your horses thirsty.

Which Lick? Minerals, Too, or Only Salt?

Veterinarians prefer pure salt licks. Salt with added minerals can prove to be problematic, as the large number of ingredients make it hard for the horse to consume the correct amount of salt. A horse's appetite for salt is naturally encoded; It's the only mineral a horse's body will immediately signal a lack of.

Many commercial feeds have salt added to them already, but the particular need of an individual horse is dependent on age, health, stress, and activity level (as well as the type of food they get). Therefore, the easiest and safest way to ensure the correct amount of salt is provided for each horse is, simply put, to give them a salt lick. Horses solve their momentary and fluctuating needs for salt instinctively—it's proven that when there's a need for salt in their diet, their taste for salt increases.

It's important to know that too much salt can be toxic, and in extreme cases, can result in death. The recommended maintenance dose for horses is 5 grams of salt per 250 pounds of the horse's body weight.

Homemade salt lick holder—be sure to drill holes in the bottom to ensure water can drain!

How to Estimate What's Lost through Sweat

If the edges of a horse's hair are noticeably white after sweat dries, you can assume he has lost a good amount of minerals. Loss from normal sweat (damp under saddle and bridle, wet neck and shoulders) is approximately 37 grams of sodium, according to the latest research. This amount is easily covered by what a horse can get from a salt lick, and there's no need to add extra salt to his feed.

In a stable, the only concern when deciding where to put a salt lick is easy access; it's usually in a wall holder or in a feed bucket. In an outdoor setting, the lick will need to be protected not only from rain, but also from ground moisture. Improvised gadgets have to provide protection from the skies and the earth.

Tip

Putting assorted kinds of salt licks in different areas with different kinds of holders increases the chance that each horse will find exactly what suits him.

Avoid Loose Salt in a Horse's Feed

Sweat contains a large amount of sodium, potassium and chlorine. Potassium is usually sufficiently recuperated through grazing and hay, while sodium and chlorine are replaced by the salt lick. But for a horse that won't use a salt lick, no matter how enticing the selection, consult a professional veterinary nutritionist; do *not* add loose salt to the horse's feed.

Professor Zeyner sums up the issue as follows: Chlorine (Cl) is absorbed faster by a horse's body than sodium (Na) is, and internal acid-base balance is therefore disrupted. This can lead to acidosis; the pH of the horse's blood decreases. In tested horses, 50 extra grams of salt could no longer be metabolized, and horses given 100 grams were clearly over-acidified. Adding loose salt to a horse's feed should be left to the experts. The rest of us should rely on providing salt licks—without consultation with a professional nutritionist, **do not add extra salt to a horse's feed***.*

The research team of Professor Dr. Annette Zeyner, who specializes in animal nutrition at Martin Luther University Halle-Wittenberg, showed in a study that **adding even small amounts of salt to feed can pose health risks***. Contrary to popular belief, scientists have found that even if a horse sweats, it's not a good idea to add additional doses of salt directly to feed. Their work shows that even small amounts can pose both short and long-term health risks: "Although access to salt is very important for horses, the loss from sweat is often overestimated. Excessive salt intake can cause long-term damage to horses."*

A self-serve mineral salt bar.

Homemade holders with canopies are very popular—you'll see them everywhere, with designs characterized by loving imagination. Owners have reported that their horses not only have preferences for which type of lick is a favorite, but also for different locations. Some horses have pretty strong preferences for which kind of holder or which site they want to use. Stumps, trunks, or logs can easily be used as holders or hangers—just carve the necessary groove or hole. It's also important to drill holes for drainage. Owners have built all kinds of shelters and covers to prevent their salt licks from swimming in puddles.

"Self-Service Bar" for Minerals

This option is for those who believe that horses will recognize a lack of not only salt, but other individual minerals. It's set up a bit like a buffet, so horses can find and replenish as needed. For example:

- **red clay**
- **dolomite**
- **diatomaceous earth**
- **seaweed flour**
- **various herbs**
- **copper sulfate**
- **salt**
- **green clay**
- **yellow clay**
- **sulfur (primarily as MSM)**

The idea of this mineral self-serve bar as a replacement for a traditional vitamin-mineral supplement can seem very enticing to those of us who want to keep our horses "naturally," but most veterinarians aren't enthusiastic about it—not only because these often use materials that, according to doctors, are potentially life-threatening, but also because this idea relies too heavily on trusting that a horse will have an exact understanding of his own needs, which isn't always justified.

Their rationale can be summed up as follows: The vast majority of horses, starting at birth, are kept in areas that are far too small to allow them to develop the necessary innate instincts to search their environment for the nutrients they need.

How many domestic horses are not only reared on sufficiently varied terrain, but then also find themselves in a new home with a biologically diverse pasture that's big enough to let them further develop and refine the experiences passed on from their dam and the herd?

If you decide to use the mineral bar approach, never lose sight of the fact that one horse's individual tastes may cause him to consume a harmful amount of one mineral that he finds irresistible. Most people I've spoken to who use these stations are happy with them and haven't had any problems yet, but here and there, complications with individual horses have been reported.

The justification behind the idea of the mineral bar is the argument that horses in the wild will choose to eat mineral-rich dirt, but this theory has only been supported by a few isolated observations. In addition, it hasn't been verified that horses even accept clay as a stomach buffer for acid—a complex mineral in response to the supposed need for a single lacking element. Recently, there's been talk of zinc deficiency. A deficit of this mineral supposedly leads to an increased intake of clay (and this very element is missing from our mineral bar). Add to that the question of quantity: how much clay does a horse really need to offset whatever he's lacking?

Minerals and trace elements in soil and rocks are present in chemical bonds which aren't easily exploitable for our horses. The better choice is always good pasture—a variety of greens in which the necessary elements are present in sufficient concentrations and are biologically accessible, with additional phyto-benefits.

Another issue I take with this system of mineral distribution is that frequently deficient elements (such as selenium, zinc, manganese, and phosphorus) often aren't even on offer.

Still, this concept continues to grow in popularity with horse owners and caretakers.

If you decide on a mineral bar, be vigilant in your initial monitoring of each horse's intake of individual components.

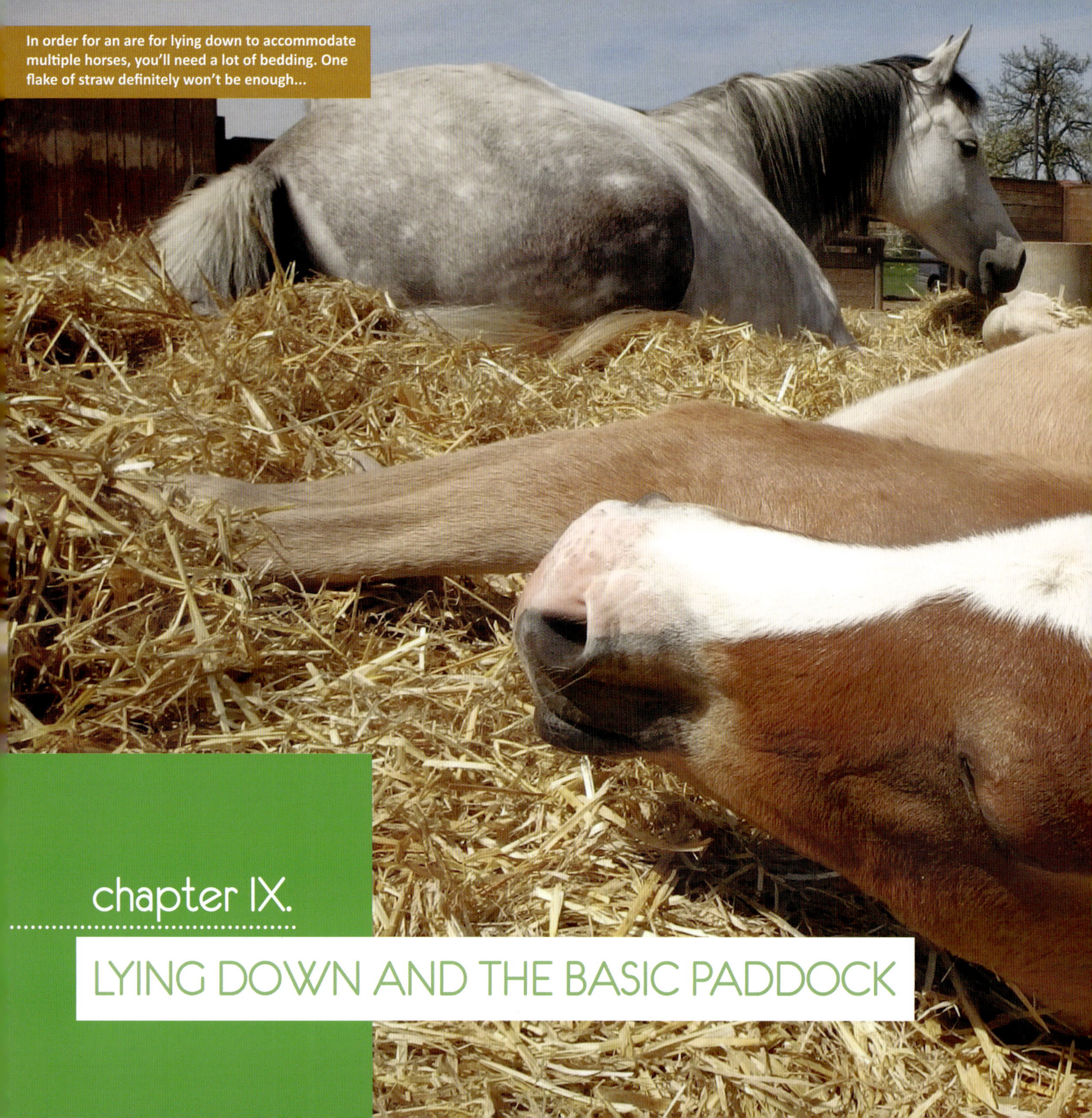

In order for an are for lying down to accommodate multiple horses, you'll need a lot of bedding. One flake of straw definitely won't be enough...

chapter IX.

LYING DOWN AND THE BASIC PADDOCK

Many horses prefer to rest on long, lush grass, but often grassy areas are either grazed down to the dirt, or sun-dried to "concrete" and mostly lacking in vegetation.

Often, horses will lie down far less often if their only options are hard-packed, rock-like surfaces.

So owners of basic paddocks add special sleeping areas, with the hope that their horses will lie down as often as they would like, and as comfortably as possible.

Adult horses only need about three hours of sleep during a 24-hour period, and it can be fragmented. Foals, however, need to sleep significantly more—including more REM sleep.

When creating a space for horses to lie down in an outdoor stabling situation, we run into one specific issue: the wishes of the owners of the horses versus the limitations of the stable and its operators. The main (and very difficult!) chore: trying to maintain a tidy space that horses lie on, play in, poop on, run through... Anyone who has tried to maintain an outdoor bed, keeping the bedding clean and in one place, will surely agree with me. Keeping such a space functional and pleasing (not only to the eye, but to the nose as well) is no small feat! So it's no surprise when some handlers answer the question this way: if your horse needs a fluffy bed, keep him in a stall!

Young foals sleep, in total, more than half the day.

Nonetheless, it's possible to create comfortable outdoor spaces where our horses can lie down.

Horse beds should be built:

- On flat surfaces. It can be inconvenient or even risky for horses that like to roll in their bedding to try it on a hill; plus, bedding gets spread far and wide.
- On dry surfaces—not necessarily under a roof, but definitely on a well-draining surfacw. If the bed gets rained on, the horses will lie down less or skip lying down altogether, depending on the material used, until at least the top layer of the bedding is dry.
- Large enough for the given number of horses. It should be possible for several animals to be able to lie down at the same time.
- Easy to maintain: easy to reach with a wheelbarrow, and with enough space to allow equipment to drive up to it and turn around.

Lying Down under a Roof

Most resting areas are under roofs. Keep in mind, where there's a layer of straw or shavings, that's where horses will prefer to lie down... and also where they'll prefer to pee. If you've built a "toilet" outside the covered space in the basic paddock, the problem will be somewhat resolved, but not entirely. Horses don't worry about whether a space with absorbent material is for lying down or for urination—they normally combine the two. In general, habit plays a role: wherever horses like to pee, they'll also like to "hang around." So, when you're figuring out where to put your horses' slumber pad, keep it away from water sources and wells that your horse shouldn't be contaminating.

It's in a horse's nature to be attracted to any soft material: hay, straw, shavings, dry leaves, sawdust, bark, wood chips, sand, fine gravel, piles of shredded paper—or anything else they can easily roll in.

Don't expect miracles: your horses definitely will pee on material intended for use as a bed.

Lying Down in Sheds

I already discussed the issue of lying down under shelter in chapter 3—now we'll take a closer look at the advantages and disadvantages.

An absolutely clear advantage is the fact that the life of the bedding is extended by protecting it under a roof, saving materials and work. For this reason alone, most people won't even consider any other options.

Frequently Asked Question:

We usually give our horses a bale of straw in the paddock, which they love to lie in. When it rains, we'd love to be able to make it more comfortable for them to lie under shelter, but we can't have bedding there. We have a horse with breathing issues who likes to stay there, and the dust from the straw in an enclosed space makes him cough. Also, whatever we put down, the horses pee and poop on. Is there any way to encourage them to use the shelter in the rain without turning it into a toilet?

Yes, there is. Either by using special mattresses (an expensive choice) or rubber load-bearing mats (a far more economical option), you can provide your horses with a straw-free but soft option for rainy and windy days. Most horses won't urinate on a solid surface—although it can happen, unfortunately, especially if one horse is persistent about it and others begin to copy the behavior. It usually will start with prolonged bad weather, making the horses unwilling to leave their shelter.

Another option is a covered bed without walls—essentially, a pergola over the rest area. The advantage is better air circulation, and thanks to the lack of solid walls, no danger of a horse getting cast.

Horses don't worry about your money. Dismantling a bale of hay that is unprotected by a feeder and lying down in it (plus other activities) takes just a moment.

You can get mattresses from dealers who specialize in farm equipment specifically for raising free-range cattle. You'll also find them abroad, designed specifically for horses in open stabling.

Here's something to ponder: One of our geldings has gotten used to peeing on the solid parts of the basic paddock, and it doesn't look like he's going to change his ways, even though he splashes himself quite a bit—the other horses even stand back so they aren't sprayed. When I started noticing this behavior initially, he was obviously bothered by the splashing and tried to lift at least one front leg to dodge the spray. Even though he has access to a large area with straw as well as a sandy spot, he didn't stop doing this. On the contrary, over time he even stopped dodging the urine, and now doesn't seem to care if he sprays his legs or not...

Of course, the disadvantages are the elements: the wind, which can easily carry off the bedding, and the rain, which easily gets under the roof.

Advantages of creating a sheltered space to lie down:

- The bedding is protected from rain and snow.
- Wind doesn't spread everything around.
- Animals moving around don't scatter the bedding far and wide.
- The horses are protected from the elements while they're resting.

Disadvantages of creating a sheltered space to lie down:

- An enclosed space is dustier.
- More stress is placed on the flooring material (dirt or clay underneath the bedding will soften from urine, for example, creating significant divots over time; sooner or later, the condition will have deteriorated enough to need repairing).
- There is risk of a horse getting cast.
- The horse has a limited view of his surroundings.

Open-air "rest area" surfaces include:

- grass
- clay
- clay with bedding
- soft rubber mats or special "horse mattresses"

Straw

Of all the possible materials you can use for bedding, I'd like to discuss straw.

Spreading a bale of straw in your basic paddock can be a good option for those who prefer it. Most commercial stables won't offer this, for reasons I've already discussed. Instead, let's look at the reasons people choose to offer this little luxury to their horses.

Advantages of straw bedding in the open air:

- There is virtually unlimited room to lie down for any number of horses.
- The air quality is better for the horses while lying down, compared to an enclosed space.

A partially covered small bed, or an airy bedded shelter? Horses don't care what we decide to call a space; what's important for them is an "affinity" for or habituation to a given area. Feeling safe is far more important than size.

- The horses will have an unobstructed view of their surroundings.
- The bedding area is easier to access with tractors and other machinery.
- There will be less odor.
- Maintenance is easy (in a dry climate, at least).

Why Choose Straw?

In the past, straw was a natural choice for a self-sufficient farm: it was necessary to bed the animals —> after grain was harvested, straw remained —> straw was used as bedding for farm animals and subsequently as fertilizer in the fields —> the fertilized fields bore good crops —> a farmer's family could live well as the result of judicious management.

What About the Modern Day?

More and more people are beginning to see horses as more than a hobby. They represent a way of life that is connected with nature and with the legacy of our ancestors. Horse owners want to be stewards for their land, managing it with respect for the natural cycle rather than just taking from it. Their animals provide them with the most valuable ingredient for their fields: manure. If you keep your horses outside, don't think of straw added for bedding as the addition of unnecessary work. No other bedding material, when combined with manure, can produce the same high-quality compost as straw. Straw can also be used to mulch vegetable gardens or layered into raised beds.

Bringing in straw for bedding with a wagon—the horses spread the bedding and after two or three days, I'll sometimes contain it to a smaller area, so it's not as prone to being scattered. In dry weather, I'll clean it daily much like I would a stall. In wet or snowy weather, I'll pick the manure and add dry straw. In the winter, I'll take out everything about once per month and start over fresh. I prefer large round bales to loose straw.

Straw is simply a fantastic source of fertilizer as well as a good addition to raised beds.

Lying down in the open air is as natural to horses as breathing. Unfortunately, most of today's horses were raised in stables, so many of them have become used to waiting to lie down until they're back in their stalls for the night, and they'll lie down only briefly or not at all when on pasture.

The more space your horses have to lie down under cover, the more likely they are to use it. If this ample space is combined with good visibility, it's guaranteed to be popular. On the other hand, a cramped covered area will likely not be used much at all. There are several factors which will determine whether the horses will use this bedding a little, a lot, or not at all.

What Can Help?

- A sense of security within the herd—harmonious interplay between individuals living together.
- Comfortable bedding, either out in the paddock or in a shelter.
- Behavior modeling—a horse who initially isn't comfortable lying down may begin to imitate the behavior of the rest of the group, if they are all happily lying down outside.
- As much adaptation as possible of your outdoor spaces to the particular needs of your horses.
- Time!

REM sleep is the stage of sleep characterized by increased brain activity, REM (rapid eye movements), irregular breathing, accelerated heart rate, and complete muscle relaxation (with occasional jerky movement). Horses will even occasionally indicate with their limbs a canter or gallop lead—at this stage, like humans, they apparently can have very vivid dreams. REM sleep is essential for brain and body relaxation.

Consequences of Insufficient Rest

In an effort to make the most used areas of the paddock mud-free by reinforcing the footing, the less-experienced owner may forget to provide a comfortably soft area that will invite the horses to lie down. If the pastures have to be closed off and the entire basic paddock is completely solid, where can the horses lie down comfortably?

Horses love to lie down in sand, it's relatively easy to remove manure from it, and sand won't affect future composting of the poop—which can be later returned to the pastures.

Horses are able to rest and sleep successfully while standing up. Sleeping while standing, with one hind leg relaxed, is something horses will do several times each day; but this kind of sleep can't completely replace the need to rest lying down.

At least once every three days or so, horses *need* to lie down to rest. The mechanism that equips the horse's body to rest while standing can work reliably even in very deep sleep; it locks the horse's limbs in a position that guarantees stability, with one leg resting on the toe of the hoof and the remaining legs firmly aligned. After some time, the hind legs will alternate while the front legs stay firmly aligned. This requires no conscious effort from the horse; he can fall into this position automatically, allowing him to spend a minimum amount of energy. Although this can provide sufficient rest, this position doesn't allow the horse to enter the REM phase of sleep. And without REM sleep at least once every few days, fatigue will begin to show.

In the absence of rest lying down, we are looking at **sleep deprivation**: the lack of full-fledged sleep with a REM phase. The reasons for this can differ—reluctance to lie down on a hard surface, a lack of a sense of security in the herd, health conditions which result in physiological problems, or ordinary old age...

Experiments on animals have shown that cats die without sleep after about 15 days; with rats, the strongest lasted 30 days before fatal exhaustion took its toll. I mention this only to illustrate the huge importance of good rest: not allowing animals long-term, quality sleep can become a serious health hazard.

Luckily, our horses are naturally equipped to handle periods when it's simply not possible to lie down. They can get enough rest from short moments of sleep while standing. Of course, as I've already mentioned, this can't replace the need for deeper rest every few days to allow regeneration of cells.

To get a good idea how much time each horse spends sleeping both standing up and lying down, a good amount of time needs to be spent around the horses each day, and at different times of day. If you take this job seriously, you'll eventually be able to notice fairly quickly if, for any reason, one of your horses is less interested in lying down than he usually is. If you notice abrasions (or worse—blood) on your horses' front fetlocks, you should seriously consider whether they have enough (or any) options for soft places to lie down in their paddock when pasture isn't available to them.

It's also important to notice how each horse interacts with the herd and how the rest behave towards each individual: if one horse is being disturbed or picked on when he lies down, a solution needs to be found immediately.

With older horses, it's surprisingly common to have deterioration of the locking mechanism that allows horses to sleep standing up. These horses can sometimes lose their balance for a moment, but they can also go down onto their knees. If this happens on a hard surface, it can result in scraped-up metatarsal joints, or even more serious injuries as older horses become more prone to falling. Older horses kept out with straw or on grass are far less likely to get injured, which is why a pasture with good turf is one of the best options for an aging horse.

Cataplexy: a sudden loss of muscle tension leading to a fall without a related loss of consciousness.
Narcolepsy: an acute pathological need for sleep which can cause cataplexy (narcoleptic seizures), although it doesn't always. Although a rare disorder, it has been recorded in horses.

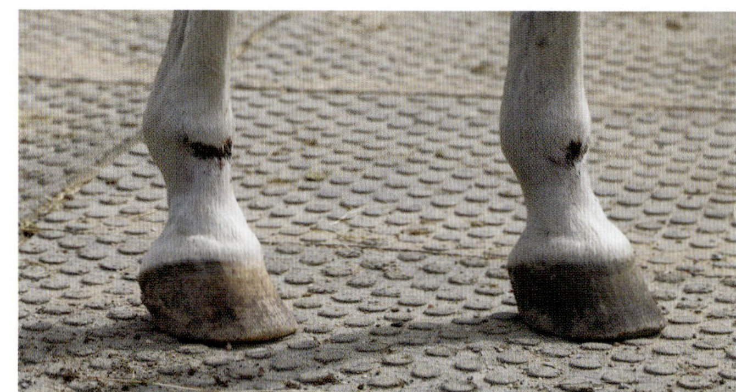

Generally speaking, abrasions on the front of the fetlocks indicate that a horse is lying down on hard surfaces, but it's not always about sleep deprivation. Sometimes horses enjoy a good roll on rougher, compacted surfaces—which, although they can enjoy a good scratch this way, can also cause rubs on their joints.

Narcolepsy has been easily identified even in the youngest foals—they'll suddenly lose coordination and fall down in front (go "to their knees"). In adult horses, symptoms under saddle have included sudden stumbles, or even a fall during a severe seizure. Some affected horses will try to overcome these attacks by trotting out. A rider can wake them with a strong leg and loud voice—asking for a quicker tempo can also help. Horses who are affected by narcolepsy need movement, and forced standing should be avoided; the advantages of outdoor living are obvious. Emotions when being taken out of a stall have been known to trigger a narcoleptic episode, whereas taking horses out of their pasture doesn't have a similar effect. (Information provided by EvaLudvikova DVM, who has worked on several cases in the Czech Republic.)

Try to see your basic paddock from the perspective of your horses. Ask yourself:

- If locked in, would they have somewhere soft to comfortably lie down, spread out, and really rest?
- Can they lie down comfortably in any position they choose?
- Can a horse lie down with a friend? Is there room for more than one horse?
- Do the horses already seek this space out?

If you answered "yes" to all of the above, congratulations! Your basic paddock will meet your horse's needs so he can enjoy quality rest.

Horses love lying down in sand. It's also easy to pick manure out of sand, and its composition isn't negatively affected by sand, so it can be returned to pastures as compost.

chapter X.

FENCING SYSTEMS

The basic paddock or paddocks must have easy--to-manage entrances which lead to pastures. According to the season, the weather, or other needs, entrances should be easily opened (for access to water, feed, salt licks, and shade), and easily closed (so horses can be kept out of an area which needs to be rested or repaired).

The Basic Paddock: Permanent Fencing

Installing solid fencing requires a lot of time and a lot of material. But if we build solid fences around the entire basic paddock, preferably combined with electric fencing, we'll sleep much better. Ensuring that horses can't easily escape isn't just a fundamental duty of anyone keeping horses, it's also crucial for our peace of mind.

Wood is the most popular option for permanent fencing. Which wood you choose will depend not just on price and availability, but also durability: acacia and oak last the longest, with an approximate lifespan of 25 years. Acacia lasts longer in dry conditions, and oak is unbeatable for wet soil.

Plastic fencing can be expensive, and the lifespan of the product can vary depending on the material. Some plastics can become very brittle when exposed to sun, while other manufacturers guarantee a long working lifespan. Always check that the fencing is suitable for horses—some kinds of plastics can break into sharp, solid fragments which have no place near horses.

Metal fencing can also be an expensive option, but its durability is unrivaled. Most fencing systems installed these days are built with electric fencing—conductors, guided along the beams by insulators. This electric addition not only serves to protect the fence from damage from chewing, but also from horses breaking beams or moving posts by rubbing. They learn very quickly not to touch the fence, making the whole area much safer. Electric fencing can also help keep out unwanted visitors—animal or human—if both sides of the beams are electrified.

If electrical conductors are combined with metal fencing, it's very important to use insulators that can keep the tape or cable far enough away from the metal in order to prevent an excessive charge.

Fencing for agricultural and forestry purposes that doesn't have an underlayment, is no higher than six feet high, and doesn't border a public space usually doesn't require a building permit. Check your local regulations to learn the requirements in your area.

Every construction project located in a buildable or already built-up area must be done in accordance with the local municipality's applicable regulations.

Barbed wire is entirely unsuitable for use around horses—it can cause serious injuries.

Enclosing a large pasture entirely with wood fencing can be unpopular not only because of the cost, but also, practically speaking, because of the time it demands. Building miles of fencing out of wood, when compared to the speed and ease of using electric tape and fencing rods, becomes a very unappealing choice. Also, keep in mind that installing permanent fencing is typically only allowed on privately owned land; if the land is leased, usually only temporary fencing is allowed.

Because of this, most people decide to go with the simpler, cheaper (although still reliable) option of the electric fence.

Electric Fencing

Tape

- Better visibility for the horses is a plus.
- The larger surface area can be a disadvantage in windy areas as it can tear in high winds.
- A strong wind can also knock down an entire line of fencing if the rods are not sturdy enough.

Rope

- Although it isn't as highly visible, it can withstand winds without problems.

***Never* install barbed wire on fences,** as the threat of injury is very real with unpredictable animals like horses. Cattle, by contrast, behave fairly prudently if caught in barbed wire; but horses will often react to pain by panicking violently, which can cause serious and even life-threatening injuries.

It's not just about your horse's safety; depending on where you are in the United States, it may be against the law to "carelessly expose" animals—or people—to injury with barbed wire, and you may also be liable for any injury to *other* people's animals, if they are wandering loose and end up caught in your barbed-wire fence.

All in all, better to avoid the question entirely. **Don't use barbed wire!**

Electrical Power Sources

Your fence can be connected to the main electrical grid, or to a battery, depending on what's appropriate for your situation. Whatever you choose, be sure it's able to handle the length of fence you have. A high-performance power source is ideal, as it will be far more reliable. Energy impulses are measured in joules—for horses, who are generally sensitive to an electrical current, 1 joule is usually sufficient.

For areas where it isn't possible to connect to the electrical grid, a battery is the only option. The power needed will depend on the length of the fence—most sellers will be able to give good advice on this. For a smaller space, a 9V alkaline battery may be sufficient.

Various types of cables are available in stores. This photo shows a Bayco conductor; it's a very safe type of electrical cord made from polyamide with a conductive braid.

Solid fencing combined with an electric fence—the wire against the white boards can't be seen...

This fencing, with its chaotic layout and mixed materials, isn't aesthetically pleasing. It will also be harder to maintain than a fence with neatly and evenly stretched cable or tape.

Longer fencing will need to be powered by something like a 12V car battery. Solar panels can be added to prolong the life of the battery. Today's energizers (the device that pushes the electrical energy out to create a pulse) are equipped with a lot of useful functions. For example, they can send a message or call a saved number in the event of a voltage drop below a set level. They can save money by maintaining minimum power until an animal comes into contact with the conductor. Some are equipped with solar panels which reduce (or eliminate, for shorter fencing in the summer) the need for recharging.

Watch out for thieves! Believe it or not, chargers and energizers for fencing are stolen quite often. Units with solar panels are unfortunately easy to spot even from a distance. If your unit will be visible to passers-by, you may want to consider installing a lockable cabinet which is cemented to the ground. Also, don't forget to protect it from the rain, with some kind of roof or housing.

You'll receive sound advice on which kind of power source and energizer suits your specific needs at a specialized store, depending on:

- **The length you need.** It's important to know the approximate area you'll be fencing in—the longer the intended fencing, the greater the possibility that distant stretches of pasture won't have the same "shock factor," and long fencing will also require higher quality conductors.
- **The height of vegetation.** There's a big difference between fencing that will pass through tall grass or have hedgerows encroaching on it and fencing through mown pasture with nothing touching the strips at all.
- **The type of animal** that will be contained by the fencing. Horses are more sensitive to electricity than, for example, cows—the pulse needs to be the appropriate strength.

The entire system should be built so that there's no need to be constantly turning off the power source, and having a fence that is often without electricity. If your horses are going to respect the boundaries set by the fence, the fence needs to always give the same sense of boundaries. If the current is sometimes there and sometimes not, it won't take long for the horses to start tearing the fence and breaking out. Every time I hear a complaint about someone's horses not respecting their electric fence, without fail it turns out that the fence was turned off for at least a portion of the day. Horses, when given the time and space, will begin to understand the differences when there's "less current" or "no current", and thus begins their fixation on escape behavior. All of my horses absolutely respect the boundaries set by my fencing for one simple reason: it's constantly functional, regularly inspected and maintained. Even young foals learn right away, and there's no further problem. Any horse's occasional accidental bump into the fence only serves to confirm what they already know, and there's no issue with escape attempts.

Gates

Entrances to a fenced-in area should be built with an emphasis on easy opening and closing. In the event that gates need to be left open, there should be an easy way to secure the electrical fencing.

Entrances should always be wide enough for machinery to pass through: an average tractor with a bucket loader will need about 15 feet between

Never use spring gates around horses! They can get their tail caught in it and pull the whole entrance down as they get entangled, panicked and injured.

posts. Personally, I like to leave a few extra feet in case wider machinery is ever needed—it can save a lot of headaches in the long run. Another option is to have your gate attached to a post that's easily removed if you ever need to temporarily widen the space.

Basic gates for horses and their handlers don't have to meet the same dimensions as those needed for a "main" entrance that will be used for machinery, but

Winding wheels with cable or tape. In the top photo, the owners have wrapped the plastic with wire connected to the conductor to protect the disc from being chewed.

The mare in this photo, after years without a problem, got her tail caught in the spring gate, which broke the gate. This poor girl was found a mile away from home, with injuries from the entanglement that took months to heal. Fear of the entrance, even after the spring gate was replaced by a safer alternative, lasted much longer.

take care not to make the gate between your basic enclosure and your pasture too narrow as there's one pretty important disadvantage: horses trotting or cantering excitedly to the pasture may bump into the sides of a gate that is just big enough. They can start breaking the posts or tearing the tape; it may be easier to build entrances generously.

A winding mechanism can be a great option at an entrance—they don't hang after releasing them, and they won't accidentally touch the ground. Even if the gate is "hot," they can be easily opened. The only drawback is that they can cost quite a bit more than an ordinary cable with a handle.

If you decide to opt for the wind-free option, tape or cable plus a handle will only cost a few dollars. This one-time savings in price, however, is paid for with greater difficulty in day-to-day handling.

The use of electrically conductive rubber has proven to be ideal for gates, and the price is becoming more reasonable. The main drawback is that it will stretch and sag after one or two seasons, which can be corrected by shortening as needed (either by cutting or tying a knot). Conductivity is retained, despite the pulling, for years.

Pasture Fencing

In general, it can be said that when we build fences on our own land, and we expect to be using them for years to come, it pays to invest in solid fencing. We can combine it with electric fencing to protect it from chewing, scratching horses. On leased land or areas that are used seasonally, as well as anywhere where there's a need to temporarily divide land, electric fencing is ideal. All kinds of fencing can be found if you look for it—you'll find fencing made from plastic spikes, posts made from recycled plastic, batten boards, poles made from acacia and oak (the most durable), spruce and cedar (shorter life span), or home-made metal rods (pay close attention to safety here; do *not* leave sharp tips exposed).

Focus on the corners of the pastures—these posts will take the most stress as the wire runs in different directions, and these posts are tasked with holding the shape and stability of the fence. Try to make corners very sturdy by burying stakes as deeply as possible so they can't tilt. When using plastic rods, it can be a good idea to reinforce the corners with at least one wooden stake or a larger diameter plastic stake.

When using wider electric tape, keep in mind that wind can catch the tape. Over time this can cause the stakes to lean and the tape will sag, which can in turn weaken the electrical charge. A very strong wind may even pull the stakes out of the ground completely. A good rule of thumb: as electrical tape gets wider, the stakes holding it should be thicker and more securely installed.

There are a wide variety of conductive tape and cable on the market; it's a good idea to consult with a dealer who specializes in electric fencing and can help you choose the most suitable options for your particular needs. Recently, I brought home Bayco wire electric fence, which so far I'm very happy with. It doesn't pull or tear in the wind, and it's flexible enough that it won't sag and still looks tidy.

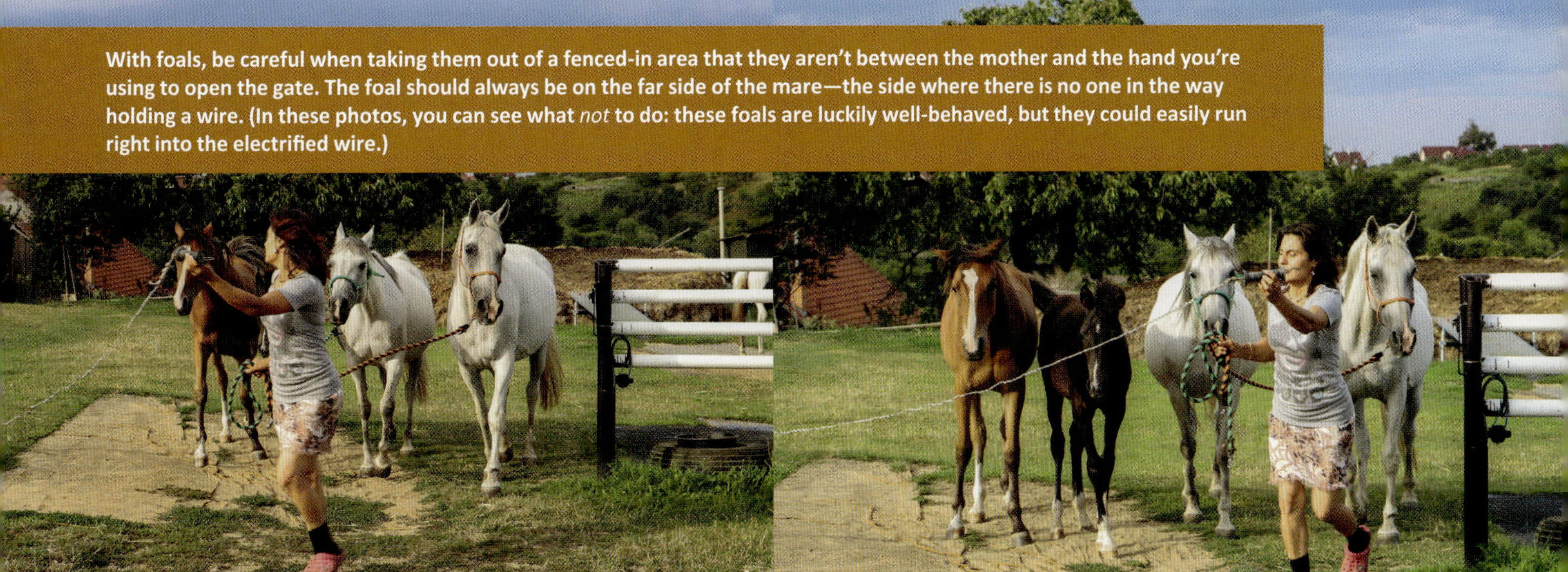

With foals, be careful when taking them out of a fenced-in area that they aren't between the mother and the hand you're using to open the gate. The foal should always be on the far side of the mare—the side where there is no one in the way holding a wire. (In these photos, you can see what *not* to do: these foals are luckily well-behaved, but they could easily run right into the electrified wire.)

chapter XI.

PASTURES—MANAGEMENT AND GRASSES

The essence of good pasture management is to take care not to overtax any one spot with too much grazing or too many animals. Once an area is overused, whether it's too many animals or too much time in one pasture, the turf is damaged beyond self-renewal. Weeds immediately take hold in available spaces, and the entire area (hilly spots in particular) is susceptible to erosion.

Unlike paddocks, pastures serve one main purpose: they satisfy our horses' need to move and graze. Anyone who has ever spent much time watching a grazing horse can't help but notice that they don't stay in one place for very long. They're constantly on the move, step by step, with their head to the ground seeking out the next bite. Unfortunately for us, this constant and often considerable physical activity (especially with young horses) which is associated with grazing, means we have to manage fields which get dug up and compacted.

Rotational and Continuous Grazing

Good and efficient use of pasture should take into consideration not just the horses, but also the condition of the fields—this is essentially good housekeeping, or rather good horse keeping. The best horse keepers need to also be experts in the science of soil management and crop production if they hope to achieve and maintain lush pastures for grazing.

Frequently Asked Question:
Should I let my horses use the entire pasture for the whole season, or is it better to divide the space and only allow the horses on one section at a time? Which is better—to graze the entire area continuously or to rest sections of the field?

It depends on the size of the pasture as well as the number of horses. Personally, I prefer to rotate and rest. Because I leave some pastures for hay, I won't use them for grazing until after the first (or sometimes second) mowing. Thanks to this style of management, areas have more time to regenerate, and vegetation thrives.

Continuous grazing is the continuous use of one pasture for grazing animals. **Rotational grazing** is the practice of dividing pasture into sections and moving grazing animals from one fenced-in section to another (and another, and so on) so that each area has time to regenerate and regrow sufficiently for additional grazing. The number of fenced areas will depend on the space available, the number of horses, and the rate of regrowth (regrowth will depend on the time of year, the climate, and the condition of the pasture: in general, the yield in most fields decreases with the number of years used for grazing).

Continuous Grazing

During continuous grazing, if climate conditions allow for continuous growth and re-seeding, the predominance of new growth (especially after periods of drought and subsequent rain) will result in an increase of nitrogen and a decrease of fiber. This can result in cases of diarrhea not just at the

When pastures are rotated and horses don't graze continuously on all of them year-round, it's easy to see the positive effects on the condition of the fields.

beginning of the grazing season, but also when the horses have already been acclimated to grazing. We can help compensate for the lack of fiber by giving the horses easy access to hay to prevent these digestive difficulties. If the horses refuse the hay in favor of grass, close them in the basic paddocks once or twice per day for a few hours with the hay. This should easily solve the problem.

Regular removal of manure in regularly used areas is a must!

Another disadvantage of having a single continuously grazed area is the increasing pressure of infectious parasites from manure that is left and the simultaneous increase in damaged areas (in many cases, this damage outright kills grasses—this isn't just sad to look at, but can also have an adverse effect on the horses' condition).

Rotational Grazing

These areas can be picked after the horses have been moved to another section (provided that the horses aren't left in one spot for too long). Before manure has been left in every spot, the horses are moved—this avoids the onslaught of parasites that would occur without the regular removal of manure.

Extensive and Intensive Grazing

Intensive Grazing

In this situation, animals are grazing in smaller areas. Pasture load is expressed in livestock units, or LSU, of 500 kg of live weight per every 2.5 acres, regardless of the animal species. The pasture load of the area in question should be managed carefully to provide enough forage. The area should be sown with grasses that are high-yield, and it should also be carefully fertilized to avoid degradation of the pasture or a dramatic decrease in yield.

Extensive Grazing

This is a large area for a smaller number of horses. This land isn't intensively farmed, and the growth and yield will therefore not be as great. Extensive surfaces, being less likely to be fertilized, won't have the same plant composition. It will be plants that are less demanding of nutrients, and more diverse. Whether or not the composition of plants growing will be suitable for horses will depend greatly on the locality (it may contain problematic species of plants).

For smaller pastures used in rotation, manure collection is the best option (if horses are to return within a few weeks). If the manure is dragged rather than scooped, keep in mind that parasite larvae will survive in temperatures below 86 degrees Fahrenheit. If surface temperatures are below this limit, dragging will simply spread any larvae in the manure over the entire area. When the grass has grown and the horses return, the risk of ingesting parasites increases significantly.

Of all domesticated pasture-raised species, horses are by far the most active. They walk continuously and can also tend to suddenly go for a run. Even more demanding of the land are foals and young horses.

This corridor leads from the basic paddock to a pasture, and encircles another pasture. From this corridor, the horses can be transferred to one grazing area or another as needed. During muddy periods, the horses don't have access to this entry, and are kept in the basic paddock. Their owners clean manure here daily.

Intensive grazing load: 1.5–3 LSU/2.5 acres
Extensive grazing load: 0.5–1 LSU/2.5 acres
(moderately intensive grazing is within the range of the given values)

Winter Grazing

During the winter, horses can be left on extensively grazed pastures if the areas are in good condition and there's sufficient space. Another option is to provide them with pasture that is intended specifically for the winter season—mowed in August, perhaps, and then left to grow. The base should be sandy or stony soil which doesn't retain moisture and dries quickly. Clay soil can be deeply damaged by horses' movement if there are enough warm rainy days; even after seeding and fertilizing in the spring it won't be able to sustain the same growth as before. It may be damaged enough that in the spring, weeds will take hold and choke out any grass which tries to gain a foothold. Remember, if you're leasing the land, after a winter with less freeze and more rain, the owners won't be happy with the state of their fields they entrusted to you.

Another consideration: grazing your horses during a wet winter can mean your horses may be exposed to grasses which have become host to fungal diseases. This can affect your horses' health if it's widespread.

Adequate pasture care and proper fertilization of land can help to prevent this; healthy grass will overwinter far more successfully! Take care, however, to avoid over-fertilizing with nitrogen—this can weaken the grass. On the other hand, a nitrogen deficiency will slow down growth and increase the storage of sugars, which can cause metabolic issues or even seizures.

Suitable types of grasses for winter pastures are those that create a firm sod and regenerate easily, can grow well even during colder temperatures, and are less susceptible to fungus or freezing.

Winter Shelter in a Basic Paddock

It's important to provide farm animals that aren't kept in buildings with reasonable protection from adverse weather conditions, predators, and other risks to their health.

This doesn't require that horses be stabled in the winter—it's enough to build a simple structure large enough for all the animals to seek shelter if they feel the need. Another option is to provide an area with a windbreak that offers relief from the elements, as long as this isn't the area where they live at all times; if it is, a shelter is still mandatory. Whatever it is, it should provide protection not only from wind, rain, and frost but also from heat and excessive sun, by giving shade.

Of course, the basic paddock as I've laid it out more than meets these requirements, but what about winter quarters?

Goutweed has no value for grazing. It has a long creeping root and can regrow from small fragments left in the soil. It is a hard-to-eliminate, nitrogen-loving weed that prefers moist, semi-shaded areas. In small quantities, it has healing effects (it's high in vitamin C, relieves pain associated with gout and rheumatism, relaxes spasms, and has a strong diuretic effect) but is unsuitable for grazing when it overtakes an area. Horses will only choose to eat small amounts and will only graze on it when there are no other alternatives. This weed drains nutrients from the soil to the point where it will out-compete grasses we would prefer to have in our fields.

Winter Quarters

This is space set aside for livestock where they can wait out the winter off-grazing season. It can be a fortified basic paddock, or a "sacrificed" pasture which can be rested in the spring. It doesn't need to have buildings or fortifications if it has good tree cover. The best choice is the smallest pasture with the least amount of grass, or a pasture close to the main buildings and therefore also near a water source and electricity.

Winter quarters for cattle.

If winter pasture is overloaded and damaged, its condition can be improved in the spring with seeding, careful mulching, and tilling. If areas need to be restored that have an excess of manure—nitrogen—then by sowing nitrogen-demanding grasses, you can prevent leaching of nitrates and other compounds from over-fertilized areas.

Bedding—unconsumed forage mixed with manure and urine left in an area over the winter that isn't used by any farm animals—is considered manure storage; laws regarding the fertilization of agricultural land usually don't apply here.

The Pasture's Shape

For our horses, the pasture isn't just for grazing, it's also a kind of playground. Essentially, it's where all their activities take place. Although a horse who lives out 24/7 won't always have extra energy to burn (when compared to horses who are stabled overnight) and will usually move at a relaxed pace, they will from time to time want to go for a gallop. For example, if they're let out of their basic paddock into their field, there will be some horses that will probably want to go for a giddy-up before settling in to graze. Because of this, it's important to know that long fence lines encourage them to pick up speed. If this long fence line ends in a corner, horses will stop suddenly, which can damage the turf (and risk damaging the fencing) and cause some unnecessary exuberant behavior: kicking out, fighting, and more.

This can be avoided by creating a ring path around the pastures: if there's no end, there's no need for quick stops or hard turns, making this "infinity loop" far gentler on the horse's body. These tracks can also be put to use during muddy periods when the pasture is closed. Horses can go from the basic paddock to this corridor to move at any pace they choose. These tracks can be reinforced with sand, or (widely used in Germany) artificial grass and various types of grids which are then covered with gravel and different kinds of sand.

Overgrazing

If horses are left in pastures for too long during prolonged periods without rain, even if it's a large area, the vegetation will become stagnant. Horses must immediately be removed from the pasture (and kept off for a while) so that the grass will be able to re-grow once it rains. If horses are left on stagnated pasture, the damage can be permanent since horses tend to graze plants down right to the surface. Grass can often be grazed right down to the roots, which causes the roots to weaken as the

Corridors (or trails or alleys) must be wide enough to accommodate the number of horses. If horses can't move freely, there's a greater risk of damage to the fencing.

If this photo is of a pasture and not a grass paddock, it's time to move the horses to another area. With the grass grazed so low, the ground suffers from loosened turf and bare spots. It should be given adequate care and rest.

plant has to pull energy from them (this is why a strong root system is a must for healthy pasture).

If, as soon as the vegetation begins to grow somewhat and can begin returning nutrients to the roots, the horses are still there to graze it back down to the ground, the roots aren't given sufficient time to recover properly. If the land isn't rested (sufficient time between grazing cycles), it impacts the longevity of the plant life. This is unfortunately why, so often, a horse's turnout is just a space with weedy dirt and nothing else growing.

The horse is a selective grazer; this picky eating can significantly impact the composition of vegetation. The plants they prefer decline, while ignored plants thrive. In addition, the horse's hooves can destroy everything in his path.

Turf to Withstand Trampling

To create a grass sod which is dense and therefore also resistant to weeds, the grasses' root nodes need to be protected. Offshoot grasses are an absolutely essential tool for the creation of a resilient pasture which can withstand the constant tread of animals. Many grasses branch just above the surface, not beneath. If we allow horses enough time grazing to shorten the turf below 3 inches, the growth is no longer sufficient to protect the nodes—which are responsible for the plant's development—from a horse's hoof. Enough height is necessary to cushion the nodes, and to maintain condition which allows for quick regeneration, which in turn allows the plant to branch out, which finally creates the density we look for. A thick carpet of grass is not only beautiful, but crucial for healthy pasture. Turf that is torn up by the quick turns, stops, and starts of our horses needs to be healed quickly with offshoots from nearby growth—otherwise the vacant area will be quickly overtaken by weeds.

If a pasture is going to be able to regenerate quickly and well, so it will be perennial and consistent, the root system must be protected so it's kept in a healthy condition. The only way to achieve this is to be sure to promptly move horses out of overgrazed areas.

Watch out for St. John's Wort in your pastures. Although it's a medicinal herb, it can cause photosensitivity in horses, and it's an invasive species in the US.

Soil temperature is critical to grass growth—far more important than air temperature. The grasses in our region, in order to have a good root network and develop into quality sod, need an optimum soil temperature range from 50° to 70°F. At soil temperatures above 95°F, grass will stop growing and become dormant. Such high temperatures inhibit the growth of the roots, which in turn can't supply enough water to the plant, making evaporation greater than moisture intake. In these conditions, there's a good chance the plant will die. The depth of the root system is the only thing which will determine if it's a dry pasture or a dying one. The deeper the soil, the more moisture it will have and the cooler it will be—the more roots there are at this depth, the better the chance the plant has to withstand the more frequent temperature extremes we're seeing. The other side of the coin: the

Providing enough time for the plants to restore nutrients to the roots' reserves is the primary objective of rotational grazing. Don't be tempted to return horses to a resting area too soon.

Grass growth can vary greatly depending on soil type, grass species, climate, and degree of damage from grazing, among other factors. However, agricultural programs at universities will sometimes carry out a teaching exercise which can quantify the average increase in grass mass in a specific pasture. One such experiment on several pastures (used by horses) showed an average increase of 33 pounds of green grass mass per 2.5 acres in a 24-hour period.

higher the above ground growth of the grass, the cooler the surface soil will be. Heavily grazed grass and longer periods of heat and drought are a deadly combination for our pastures.

If drought is prolonged, not all species of grass are capable of regenerating when rain finally falls. This is another factor which can transform a valuable, lush pasture to a field full of unwanted weeds...

It's plain to see that the goal of all the effort put into maintaining pastures is an effort to create and maintain a dense sod in excellent condition: without bare spots or weeds that horses won't eat, and without weeds that could cause problems or are directly toxic. Excellent condition also requires that we don't overload the area with too many animals, and we must take care of the supply of nutrients. We also have to cycle or rest the pastures in a timely manner, and to never skip necessary steps like mulching, re-seeding thinning areas, and fertilizing according to the soil's needs.

A horse's grass consumption is around 8 percent of his weight each day. By this measure, a 1,100-pound horse can graze approximately 90 pounds of grass in 24 hours. Other data for horses grazing freely shows a lower intake—which can of course depend on the palatability of the grass and how much is available. Compare these numbers while considering the average increase in pasture mass per acre!

Dragging

Lightweight panel or chain-link gates are great for dragging pastures. The side with small spikes helps aerate the soil that's been compacted by the horses (soil compaction can limit good grass growth).

The side without spikes can spread manure and level unevenness (like, for example, a mole hole). These gates can even level holes made by horses' hooves, if the treads aren't too deep, and they rip through moss and old remains. Always drag in dry conditions to avoid leaving tracks, and be careful not to tear the sod, which can disrupt the interconnectedness of the grasses. This is a plus if the growth is overly dense, but that's rarely a problem for horse owners, unfortunately... After sowing and fertilizing, spreading is necessary. At the end of each grazing cycle (as long as the temperature is high enough or the horses won't be returning back to the same area until the next year after the first hay), it's fine to simply spread the manure. Using the smooth side of the drag, it's easy to break up manure without damaging the turf. Pay close attention to the fence line to avoid catching posts.

Frequently Asked Question:

After this winter, we have meadows full of molehills. I'm afraid the horses will be grazing on more dirt than grass. What can be done?

Moles can be a big aggravation. If horses are sharing a space with a mole, their movement will usually encourage the little guy to find a quieter area, and the problem will fix itself. But if pastures are left empty for the winter, the spring may show what these animals can do when left alone... Moles can damage sod, creating bare spaces. Also, they can cause problems with haying: hay gathered from molehill-infested areas can be excessively contaminated with dirt, not only causing problems for horses who already have respiratory issues, but making even healthy horses cough. Experience shows that having dirt mixed with hay is dangerous for hygiene and the quality of the hay being baled. Moles in pasture after winter can cause a delay, which can be corrected if there are only a few of the critters, and we don't expect high yields from the field. If the field was overrun by moles, however, especially if more grass or hay is expected from the area, it may be necessary to completely recultivate and re-seed the pasture.

Last but certainly not least, keep in mind that there's a risk that a horse can step on a molehill and fall through. Depending on how fast he's moving, this can have consequences for the horse's musculoskeletal system.

Fertilization

Why fertilize? After all, horses leave piles of poop which can be spread—instant fertilizer, right?

So, Don't Fertilize?

No, definitely fertilize! Preferably with organic fertilizer. I'm going to address the issue of fertilization purely from the perspective of supplying soil nutrients, carefully spread over the surface. I'm *not* talking about "fertilizing" as it's often done, essentially through lack of maintenance—see the photo on page 134. This neglect leads to an excessive concentration of manure, which not only kills grass, but also promotes environmental contamination by parasites. This issue is covered in various veterinary publications, so I'll focus on fertilization intended for maintenance of soil health.

Nitrogen (N) affects the quality and yield of forage most out of all the elements I'm going to discuss. It also affects the representation of species in the field. In combination with other elements (for example, phosphorus, which increases the efficiency of nitrogen fertilization), it is indispensable for growth. Viability varies—many factors play a role. In general, the carrier of most nitrogen on the soil surface is dead organic matter, but this matter must first go through the decomposition processes of smoldering and rotting. The ready availability of nitrogen is therefore very small, only about 5 percent! Although the current trend is to avoid synthetic fertilizers, this is not so simple. If we want to achieve healthy growth in an area so our animals can really graze, especially in heavily overgrazed areas, we need to help ourselves by using a reasonable supply of an inorganic fertilizer, at least in the spring.

If we combine regular organic and inorganic fertilization components, we can both support long-term microbial life in the soil and also meet the immediate demands of the plants.

Excessive supply of nitrogen from horse manure, left in piles, promotes the growth of nitrophilic flora. This results in an increase of nitrogen-loving weeds, like sorrel, swallowtail, nettle, gorse... which significantly degrades the quality of grazing areas.

An unmaintained pasture soon turns into an area full of suffocated grass, piled-up poop, persistent and hard-to-kill weeds, and parasitic infestations. Is this "natural" state worth it...?

Phosphorus (P) effectively increases the availability of nutrients for vegetation. Inorganic fertilization isn't a suitable option, because plants can't absorb it when it's in the upper layers of the soil. Only about 3 percent is released into the deeper layers, and in acidic soils, unless the pH is adjusted, phosphorus is practically unavailable. To increase the intake of phosphorus by a variety of plants, we have to change our thinking: although completely removing manure is a great idea to prevent the spread of parasites, it takes away the chance to return nutrients to the soil. Many stables have carefully cleaned paddocks and pastures—distributing the manure they collect to neighbors for their gardens and fields. The neighbor's fields then thrive, while the stable's own land is impoverished of beneficial organic material...

Those that collect manure and allow it to ferment and compost before returning it to their fields are not only managing the land for future generations (by adding indispensable humus), but also for themselves and the current quality of the grasses. This is the best way to ensure availability of phosphorus for vegetation. Sowing leguminous plants (like clover and alfalfa) can improve soil health: alfalfa (which benefits greatly from the phosphorous provided by our composted manure) has a deep root system that aerates compacted soil and takes up phosphorus from otherwise hard-to-access compounds, while clover fixes nitrogen and aids in the absorption of phosphorus needed by other plants.

Potassium (K) plays an important role in the growth and resiliency of plant tissue. Grass cover will generally have sufficient potassium, because it can draw it even from less accessible bonds in the soil. Both in grass and hay, it can often be found in excess. Due to the fact that it acts to counter magnesium, this can be a problem. Although a harvest may not need potassium fertilization in order to avoid deficiency, the soil may become depleted. Finding the right solution can be a delicate balancing act.

Fertilizer isn't applied to the soil just by our animals (manure) or by us when we fertilize or mulch, but also by Nature herself—rains and storms can fertilize, and dead plant mass turns to compost. How?

Old grass, root systems, blown leaves, and the soil's decaying dead microorganisms are helpful decomposers, transforming dead organic matter into the best possible fertilizer and providing necessary mineralization to vegetation. Also, the addition of leguminous plants in the grazing areas is an effective way to provide some of the nitrogen that grasses need. Thanks to their symbiotic bacteria, these plants fix atmospheric nitrogen in their roots. This ensures the health of soil that isn't fertilized mechanically or by manure. Leguminous plants suitable for grazing include alfalfa and some kinds of clover.

Fertilization affects the variety of plant cover, as well as the mineral content and organic substances in the vegetation.

However, it's not recommended to give horses access to too much clover or alfalfa—they shouldn't have all-day access to any field if the percentage of these plants is higher than 20 percent. Most seed mixes manufactured for horses which include alfalfa or clover take their specific digestive and nutritional requirements into account.

Rain supplies approximately 20-30 pounds of nitrogen, and some potassium is released into the soil over time as it weathers (rocks break down). But this on its own isn't enough when year after year, the pasture's yields are exhausted by grazing and trampling. Without extra fertilization and protection from the horses (rotational grazing), the pasture will soon enough be full of weeds and bare spots.

Soil is comprised of clusters of particles of various sizes, which were formed by the gradual breakdown of rocks (weathering). A number of factors affect the rate of weathering, such as rock composition or climate. This is a long-term affair—it can take hundreds of years to form one centimeter of soil, and in some cases it can take even thousands! Doesn't this make you respect your soil a bit more?

In order for this soil to then bind water, it has to be mixed with organic substances that have undergone humification processes—become *humus*.

Moisture and health are the two most essential elements in caring for our pastures. Moisture is mostly out of our control, but we can influence soil health. Organic fertilization can be used to protect grass: higher microbial activity in soil encourages the growth of plants that are more resistant to stress and disease.

In general, a field with a composition including 1 percent leguminous plants corresponds to about 6 or 7 pounds of nitrogen in mineral fertilizer. These fertilizers are great for grass, but they suppress clover. Fertilizer made from composted manure promotes more balanced growth, including for clovers.

Humus is made by soil organisms like the irreplaceable earthworm. This humble creature not only transports humus, but also helps to create it. Thanks to this humus, plants' root systems can receive oxygen. Humus is responsible for the porosity and absorbency of soil, and also carries a supply of mineral substances which, through the action of microorganisms, is gradually released and used to feed the plants.

In the case of a large area with only a few horses, fertilizing with animal manure should be completely sufficient to support plants that haven't been stressed by excessive grazing. These plants, not having had to re-grow repeatedly, won't have depleted the soil and will need less support. These lucky people won't have to resort to artificial fertilizers.

Unfortunately, in the majority of cases, this natural cycle of nutrient supply isn't enough. Areas for horses are exhausted from overuse: overgrazed, compacted, and often depleted from prior years' poor management.

Here, there's more than enough alfalfa. Horses should only be allowed in a field like this for a limited time. As the amount of alfalfa decreases, their grazing time can be proportionally increased.

Natural erosion from wind and water: in a vegetated area, the soil loss will amount to approximately 1 centimeter over the course of 20 years. On bare soil, however, this process can occur so quickly that topsoil can literally disappear right before your eyes. This is why we must be caretakers not just for our horses, but also for the land. It's our duty to maintain high-quality grass sod. This may mean prioritizing the health of a field over the immediate needs of our horses (limiting their space) when the situation requires it.

Freezing of pastures and susceptibility to winter fungi can be a sign of a pasture needing nutrients.

Some surfaces might be literally dead: unable to produce anything without the help of synthetic fertilizers. Many owners choose to just seed these areas and turn them into pastures, but soil lacking in organic material is essentially deficient, and won't be capable of supplying productive mass...

Fructans

Fructans are sugars that store carbohydrates. From the point of view of horse keeping, this information is crucial. Where soils have depleted nutrients (like nitrogen), plants will accumulate sugar reserves in their tissues until the time comes when they can use it.

Grass Stress

Unfavorable conditions that force grasses to resort to dormancy: prolonged drought (especially during hot weather), too many horses for the area (which negatively impacts load-bearing capacity of the turf and compacts the soil), grazing to the roots of the plant, and poor soil health (either lacking nutrients or poorly balanced nutrients). These conditions will stop growth, but photosynthesis continues, so the plants will continue to produce sugars, accumulating them in their tissues until the conditions are favorable for the plants to use them. This happens quickly, as plants try to survive and keep their territory.

Viper's bugloss (also known as blueweed) has become naturalized in some parts of the US but is invasive in others; it's a fantastic plant for insects, as pollinators tend to like it, but horses typically dislike it. Under some circumstances, it can be used to treat diarrhea, and as a disinfectant—but in larger quantities, it builds up in the livers of both cattle and horses, and is toxic. It was once believed to be effective in treating snake bites, but scientific tests have disproven this idea.

Stagnant growth, due to either lack of nitrogen or unfavorable weather conditions (drought, heat, cold), combined with high storage of fructans, can be problematic for most horses. Fructans will help grass survive the cold, but can have serious health effects for our horses, especially those that are insulin resistant, or prone to seizures or obesity. This is why it's in our best interest during dormant periods to not only empty the fields, but also to be sure to return fertilizer to the soil. Since autumn doses of horse manure from a few horses won't be enough to provide sufficient nutrients for a large piece of land, it may be necessary to supplement with mineral fertilizers in the spring.

Warning:

When nitrogen fertilizer is increased to encourage growth, magnesium levels decrease. When many horses stay on a small area during the winter months (a "sacrificed" area that's re-seeded in the spring), testing has shown high concentrations of nitrogen, phosphorus and potassium will build up in the soil.

If grass is deficient in nitrogen, sugars accumulate in tissue (photosynthesis slows but doesn't stop). If there's an excess of nitrogen, the tissue will mature more slowly and will be more susceptible to fungal diseases and frost.

The organisms in our soil are absolutely indispensable; without them, the transformation from organic substances to dirt wouldn't happen. Unrelenting interference from things such as industrial fertilizers, residues from medicines and dewormers, and carelessly used insecticides are literally killing the ecosystem of our pastures.

Magnesium often turns out to be lacking in horse's diets, including when horses have been over-grazing an area.

If you have a soil analysis done, you'll know what to expect. If any element is noticeably lacking in your soil, it won't be present in grass or hay that grows from that soil without targeted fertilization.

Don't underestimate the importance of fertilizing for magnesium. It's particularly important at the beginning of the season, when there's often not enough of this element to support the rapid growth of grass.

The horse in this photo is having fun, but sometimes horse-like behavior can be atypical for the horse in question. If you're constantly trying to handle a horse whose nerves seem disproportionate to the situation, it's worth checking the magnesium content in the grass and hay.

Fertilizing with magnesium is unfortunately fairly expensive, which is why it's often neglected. Testing has shown that hay from many suppliers is regularly magnesium deficient.

Magnesium Fertilizer Tip:
Dolomite limestone is an excellent all-natural magnesium and calcium fertilizer; it has a high content of calcium, magnesium and also trace elements. This should be your first choice for a gentle, long-term solution.

Be sure to allow time for the soil to absorb the nutrients before using the fields: you'll be rewarded with a healthier root system, denser turf, and more nutritious forage.

A field with moss and old, overgrown plants has probably been overused and hasn't had necessary maintenance carried out. This can be a suitable pasture for winter grazing, but it will be absolutely necessary to remove it during spring months and properly maintain the area. Overgrowth will prevent development of new plants by shading them from the spring sun and not allowing the soil to warm enough. Among this old growth will also be plenty of weeds, which, if left alone, will spread and take hold.

In the autumn, when an area is no longer being used for grazing and the remains have been mulched, it should be fertilized with composted manure. In the spring, commercial fertilizer (such as a basic NPK fertilizer) can be used as needed. If you don't want to go by guesswork, soil analysis can be done to determine what specific deficits should be targeted, or whether fertilization can be avoided altogether.

By dragging an area that's been "fertilized" by a year's worth of horse manure, you'll ensure an even layer of organic matter, including spreading mulched grass (which is mulched before fertilizing). At the same time, the drag will tear up any moss and remnants of trampled overgrowth which the mulcher may have missed and which would otherwise suffocate new growth. Any uneven or bumpy area should be dragged in several directions to avoid leaving clumps of material. Dragging doesn't have to be perfect—a season of snow and then a melt, followed by spring pasture maintenance, will hide a number of mistakes.

The gates used to drag a pasture should always be light, as a heavy or hard drag will tear up sod along with moss and weeds—leaving bare spots which give an immediate opportunity for weeds to take hold.

Soil is a renewable natural resource—but that renewal takes work. Without reserves of organic matter, it's a dead zone. Let's appreciate and use the daily supply of horse manure (through composting and applying (or directly spreading if the land will be rested).

After fertilizing, give the soil a few weeks to absorb the nutrients and for grass regeneration. You'll be rewarded with a healthy root system and thick turf that horses won't trample as easily.

Rolling

Rollers are used only after seeding, to push the seeds well into the soil. Rolling should be done when the soil is slightly damp—never when it's too wet or too dry. If there's no need to re-seed an area, there's also no need to roll it: when horses have been walking continuously in an area, it's always best to avoid additional unnecessary compaction.

In an unreinforced area of a basic paddock (for example, a bare clay section), the situation will call for rolling. The surface should be compacted, to more effectively protect it from being churned up by the horses than by simply leveling it.

Rollers are most effective when used before a deep freeze—if an area is rolled before the onset of severe frosts, it's more likely we'll be able to prevent the paddock from transitioning from muddy and churned up to frozen craters and divots. Horses don't like to walk on hard, uneven surfaces as it can be painful for their hooves—they may even avoid walking to water if the footing is bad enough. Even if the surface can't be made perfectly smooth, rolling it before it freezes too much can help compact and even out the ground enough to avoid painful, sharp ridges.

We should always give preference to "farm" fertilizers, only supplementing with industrial ones when absolutely necessary.

Depending on the weight of the roll, you may have to experiment with the best temperature for the best results (the ground needs to be somewhat frozen, but not too much).

Nature has a clever way of using weeds to fill gaps in a grass field.

Mulching

Mulching and mowing in order to prevent weeds from flowering is a necessary intervention. If weeds are allowed to go to seed, they'll replace more and more of your grass. If a weed is mowed after it blooms, it may still be able to go to seed even after being cut—this is why unwanted plants should be cut before they can flower. For an acre or two, it may be possible to address patches of weeds by hand with a sharp scythe.

Fallow spots in pastures, if left alone, become prime locations for weeds. Horses won't graze in blighted places: if an area has a good amount of manure, horses won't graze on it, even if the vegetation is allowed to keep growing. Horses prefer to keep grazing in over-grazed areas, literally down to the roots and dirt, rather than eat in the "toilets." This kind of treatment will destroy a pasture: places used only for manure will expand, and popular grazing spots will get smaller and smaller. After a few years, overgrazed areas will have suffered too much to compete with weeds, resulting in trampled dirt areas with islands of weeds. Only well-timed mulching or mowing (and possibly removing—if it's a lot of growth mixed with manure) can prevent the complete destruction of grazing areas.

When There's a Lot of Space

It can be a completely different situation if there's so much room that the horses have more than enough to graze on.

Here, we can let nature take its course with far less frequent human intervention.

It's worth noting that horses will graze more selectively in these larger spaces (our four-legged friends will take advantage of the wide selection and will be choosy about what they eat), leaving a much higher degree of pasture completely uneaten.

Extensive grazing on a large space by a small herd can eliminate the need for mowing without

A nice grass field is the best possible place for a domesticated mare to foal. Births on well-maintained pastures are becoming more prevalent.

creating damaging changes to the vegetation. If we're comfortable with a greater variation in the composition of vegetation than we might want on intensive pastures (including some grasses and herbs that provide lower yields, plus places overgrown with taller woody plants), it's a fantastic option for our horses. Let's keep in mind that it suits our horses to graze almost non-stop with a great degree of movement. So instead of having tasty grass everywhere they look, they'll search a diverse spectrum of plants. Our horses aren't traditional livestock: they don't have to achieve record milk yields, or gain a maximum amount of weight, so the species composition of grasses can be far different on a large area when compared to intensive grazing areas.

This kind of natural meadow for year-round grazing represents the best we can give our horses. If we add hay and feed based on the individual needs of our specific animals (not to be confused with our assumptions about their needs), we can rest assured that even during the harshest months of winter, these horses are living their best lives.

Seeding Pasture and Thickening Turf

Keep in mind that the more grass is cut before it can flower, the less it will branch off, creating thicker sod. If sparse or bare areas need to be filled, grass shouldn't be left to grow and go to seed—it should be mowed before it's grazed to support the thickening of the grasses. In practice, however, this can be unrealistic in drier regions. Especially for smaller areas with many animals, the droughts that have prevailed in recent years have made farmers hesitant to risk mowing too early or too much. If a farm produces its own hay, early mowing can lead to dried-out growth during subsequent dry spells. Even if there's rain eventually, these mowed areas will have gone dormant, leaving dry, yellow plants. In

these drought-prone regions, farmers have to rely on supplementing the growth with regular re-seeding in the fall and spring. In regions with more temperate weather, pasture management can combine methods—alternating between seeding and mowing.

In addition to creating a thick sod, this will also produce a highly nutritious hay. There won't be much at first, but the second cut will make up the difference. Don't forget that second-cut hay, although highly nutritious, is not ideal for feeding freely to every horse. It's far too rich for an easy keeper, but is ideal for horses with increased energy demands such as pregnant mares, growing young horses, thoroughbreds, and possibly older horses who struggle to maintain their weight.

Grass left intentionally to flower and go to seed (that is, significantly longer than it would be left if it were being mowed for traditional hay production) can be a helpful way to finish a pasture. The hay isn't lost, since it can be cut and dried after it drops its seeds, but it will be an energy-poor straw hay suitable only for obese horses or very easy keepers.

Seeding should be done in either spring or autumn when there's an excellent chance for light and steady rains. If seed begins to germinate but then dries out due to drought, the seed will die, and later rains won't help it. If drought conditions occur after sowing, but before the seed has germinated, it will still most likely sprout after later rainfall. Your best chance for a high germination rate in this situation is to sow pastures using a seeding machine that presses the seed into the soil (rather than by hand) to protect it from hungry birds while it waits for rain. This method ensures a germination rate as high as 80%.

Deworming and Its Effect on Organisms in Soil

According to (fairly recent) research, some deworming preparations for horses—specifically ones containing ivermectin as their active ingredient—are strongly toxic to coprophilous beetles (popularly known as dung beetles, who reintroduce nutrients from manure back into the soil). Your horse's manure will be toxic to these creatures for up to 45 days after its administration.

How can you prevent the extermination of helpful organisms like these in your pastures?
Avoid using products with ivermectin in them during the spring months. If it needs to be incorporated, add it in at the end of autumn, when the activity of these highly useful beetles is minimal. They are at the height of their productivity in the spring, and would be highly attracted to ivermectin "scented" manure, causing a die-off of these little cleaning helpers and soil fertilizers.

Coprophilous beetles
(beetles that feed on manure in their adult and larval stages)

- bugs
- dung beetles
- notched beetles

An area with extremely varied vegetation. If we were to walk through this five-acre parcel, we'd find all kinds of grasses (including ryegrass, fescue, linden, and alfalfa), legumes here and there, a little sedge, red and white clover, yarrow, plantain, dandelion, and more. Due to the large amount of alfalfa, the owners of this field are careful to gradually increase their horses' grazing time. Only after a few weeks, when the pasture is more heavily grazed, do they leave their horses here for up to a half day. They have noticed that their horses like to start with the alfalfa, and over time they add in other grasses. Clover is eaten only sparingly.

Grasses

Grasses are pollinated with the help of the wind, from this wind-borne seed a primary root sprouts. Grass seeds spread thickly but grow shallow roots—often only as deep as 6 inches below the surface. They therefore need more frequent rainfall (even if not a soaking rain). At the top of the root, either just below or above the surface (depending on the species), a branching node develops from which bundled roots and shoots grow. The blades have the highest nutritional value out of all the parts of the grass.

The value of grass is not only what it contains in terms of nutrition for our horses, but also in its ability to create a fertile and sustainable grassland for livestock.

Any surface that horses regularly use will be very compacted by the constant movement, which reduces the soil's ability to absorb sudden heavy rainfalls during dry periods. Although we need our basic paddocks to stay dry and drain well, pastures are another matter: they need to be able to retain water deep in their soil.

Our desire to maintain green space everywhere for our horses runs counter to our horses' destructive impact on growth.

If grass has a healthy root system, the soil will have invaluable biopores (created from decaying roots and soil organisms), which can allow deeper penetration of rainfall, better aeration, and therefore better root growth. The roots and biopores together maintain a healthy soil profile which enables better growth above-ground.

Roots which can grow deeper can access nutrients deeper in the soil. Deep root systems reduce erosion and preserve the quality of groundwater. A rich root system can also prevent seepage of nitrates into lower levels of the soil. Dead and decaying roots also help by enriching the organic soil profile.

Categories of Grasses

Drought-tolerant grasses:

- Tall fescue (cool season)
- Wheatgrass
- Alfalfa
- Reed fescue (very deep roots)
- Bahia grass (warm season, low forage quality)

Warm-season grasses:

- Bermudagrass
- Crabgrass

Kentucky bluegrass.

Cool-season grasses:

- Orchard grass
- Perennial ryegrass
- Kentucky bluegrass

Grasses and legumes for moist soil:

- Tall fescue
- Reed canarygrass
- Red clover
- White clover
- Birdsfoot trefoil

Grasses for areas with ideal rainfall:

- Meadowsweet (medicinal herb)
- Perennial ryegrass
- Red and meadow fescue (if endophyte-free; otherwise, it can be toxic)

Branching:

- **Intravaginal shoots:** new growth forms as an offshoot from the mother stem, resulting in an erect growth habit, very little lateral spread (plants will form bunches). Growing points are higher off the ground and are susceptible to removal by grazing.
- **Extravaginal shoots:** new growth via rhizomes or stolons, creating new growth independent of the mother stem. Sod forming with close-to-the-ground growing points, less susceptible to removal by grazing.

Growth patterns:

- **Bunch grass:** fast initial growth, high yield, limited endurance.
    - *with spreading habit:* has grazing value, large yields, and high forage quality. Meadow fescue is a good example.

- *with dense growth habit:* less valuable for pasture due to low yield and low forage quality—for example, sheep's fescue.

- **Protruding growths:** slow initial growth from seed, long-lasting, tolerate trampling, good for covering bare spots.
  - *with canopy structure* (rough bluegrass, velvety bentgrass ...)
  - *with undergrowth* (red fescue, Kentucky bluegrass, meadow foxtail, and couch grass—invasive in the US, but unfortunately common)

According to vernalization (spring germination after winter temperatures):

- **Winter grasses:** for example, perennial ryegrass—this grass blooms once a year, after passing through the "vernalization stage," or winter temperatures. Without passing through these cold temperatures, the seeds can't bloom. These grasses do most of their growing during the spring; after the second mowing, they produce far less.

Bunch grasses with dense growth habits don't tolerate compaction or trampling. Bunch grass with a spreading habit and protruding grasses not only tolerate trampling, but these stressful conditions cause them to branch out more widely.

- **spring grasses** (for example, annual ryegrass and meadow sedge: these grasses bloom several times per year and don't require cold stratification for seeding. They grow reliably throughout the growing season, and grow the best during warm days.)

Grasses for Hay

Apart from water, these grasses have only minimal requirements. Professional literature states that these grasses are "ecologically flexible." Their use as fodder can be influenced easily with agro-technology, especially the species and fertilizer requirements. The quality of the hay can also be affected by the time of harvest, how it's mowed (for example, mowers equipped with conditioners will crimp the hay and speed up drying time), and weather conditions during drying. It's ideal to harvest hay within three days of drying to avoid fermentation (the natural process of nutrient loss). The way hay is stored also has a great effect on its quality and safety.

Grasses are at their most nutrient-dense at the time of mowing. After grasses flower, the nutrient content decreases, and stalks and leaves become woody and more indigestible. Grasses with thicker stems grow longer, and generally have more fiber and lignin (which is what makes the plant cells tougher) than fine-stemmed grasses. Fine-stemmed grass is more easily digested (having less fiber and more nitrogen).

High-production grasses for hay:

- meadow fescue
- orchard grass
- timothy grass
- Kentucky bluegrass

Hay fields can benefit from additional doses of nutrients, which they convert very efficiently into new growth. Grasses such as ryegrass, ryegrass hybrids, and bluegrass require lots of nitrogen.

These species very quickly flower and go to seed when there's a lack of nitrogen, creating fewer leaves (less forage). By not fertilizing (horse manure in the autumn, NPK fertilization in the spring, and, for acidic soil, salt water), yield is reduced.

The nutritional value of grass varies based on the type, but generally speaking, short grasses produce more leaf mass and are therefore more nutrient-rich, while taller grasses have more stem than leaves and therefore have less nutritional value. There are exceptions: for example, Kentucky bluegrass has more leaves than perennial ryegrass, but its nutritional value may be lower.

The growing stage can also affect the nutritional value of grasses: the younger the plant, the better its digestibility and the higher the nutrients and minerals. Older growth is woodier (has more stems) and more indigestible fiber.

The second cut, and every subsequent harvest, has finer, shorter stalks. It has less fiber and sugar,

Hay can only contain what the grass possessed before it dried. After a few seasons, grass from areas that aren't properly maintained won't have the level of minerals normally expected for that particular species. A good knowledge of the nutrients contained in the soil of a particular location and the management of pastures and areas which are set aside for hay can be the difference between thriving horses and horses in poor condition.

and more protein. If hay is harvested in the autumn, when temperatures are cooler at night and in the mornings, the stored sugars in the grass may be higher again as growth slows and plants store more fructans.

Temperature is an important factor when it comes to the growth and mass in pastures and hay fields. During high temperatures, growth accelerates and stems will lignify (become less digestible and woody) faster.

If temperatures are high (above 85°–90°F), and the grass isn't getting enough moisture, accelerated death of the leaves begins. This results in growth which is far more straw-like, with a much lower concentration of nutrients.

Excess energy in cool-season grasses which are stored as sugar (fructans) aren't stored evenly: most of them accumulate at the roots and the base of the plant. The taller and older the stalk, the fewer fructans it contains.

If land dries out, more and more land is needed per horse. If this isn't a possibility, another approach is to use the basic paddock more actively and save pastures for limited turnout. You don't have to give your horses only hay!

Types of Grass for Pasture

Depending on what type of pasture you'll have, whether they're intensively used areas where horses will only spend part of the day, or extensive grazing where horses graze freely throughout the season or year-round. If your horses have already grazed everything there is at the beginning of the season, and the rest of the summer they're on more or less trampled areas, you'll need a very hardy variety of grasses (quick growing, tolerant of trampling) than if your pastures have a surplus of grass no matter how much it's grazed. In areas where open land is at a premium (for example—closer to cities), there will often be quite a few horses on over-grazed, dried out or muddy lots.

In areas with limited space, we have to be OK with limited grazing to preserve the space, leaving our horses in basic paddocks with hay for the rest of the time.

In areas with highly productive grasses, thanks to which there's more growth for a longer period, we can afford to leave horses to freely graze, provided that we don't leave them in the fields for too long. The amount of time spent in their basic paddock versus pasture will depend on the abundance of growth, and should be monitored.

A horse's digestive system needs time to adapt to spring grass after a winter kept on hay. The intestinal microflora need time to adapt and multiply: the new fodder (grass) will need new, healthy colonies of bacteria for digestion.

How to Acclimate a Horse to Grass

Horses that are only on hay during the winter should be put very gradually on pasture. You can start by hand grazing for 5 minutes (gradually increasing to 10) before and after they're ridden. For out of work horses, you can bring grass to them from the pasture (first a bucket, gradually increasing the amount). When we open the pasture gates, the horses will be ready for about 30 minutes of grazing before they're brought back to their paddock (some prefer to leave the horses out, and gradually increase the space for grazing by moving the fence a bit each day). The amount of time can be gradually increased every few days. By the end of three weeks, they can be grazing for a few hours at a time.

With trouble-free horses, this schedule can be sped up slightly. However, you can keep horses who need more careful monitoring (horses with a low tolerance for sugar) on limited grazing throughout the season. If all goes well and none of the horses have loose manure, after a month you should be able to leave your horses on a well-grown pasture for half a day. If the pasture is poor and over-grazed, they can be left even longer. Within the next two weeks, conditions allowing (acreage, type and growth of grass, weather, breed and health history of horses), they can be allowed on pasture for the entire day. In some locations, this schedule can be sped up a bit, but keep a close eye on the consistency of the manure. Some horses (those with metabolic issues) won't be able to handle all-day grazing even when it's introduced very carefully. With basic paddocks, there shouldn't be any problem limiting grazing time. I recommend fencing off some smaller areas in the pasture to make it easier to catch horses who will need to come off grass earlier.

All-day grazing for our horses can be allowed in productive pastures either with the system of gradually moving the fence line to add fresh grazing to what's already been eaten down on previous days, or by making the entire area available for the whole day only when the pasture is grazed down quite a bit. However, this doesn't mean we should let vegetation get destroyed unnecessarily—grazing should be stopped and fields rested whenever necessary.

If you're lucky enough to have lots of space and only a few horses, the grass species chosen should be low in sugars. Most specialized retailers will be able to give advice as to what seed would work best for your location. If we allow horses unfettered access to plenty of space sown with high-yield grasses (for example, the type used to graze cattle who are being raised for milk or meat), the next call will be to the vet...

Seed Suitable for Horses

When choosing seed for establishing or restoring a pasture which will be used for extensive grazing for horses, it's not only important to know which grasses are specifically recommended for horses, but also which specific mixes will thrive together in a specific location with specific conditions.

Horses on pasture year-round are naturally acclimated to new grass—they will slowly nip away at more and more new growth along with winter's old straw until spring's fresh growth is established.

This is when it'll serve you well to consult with a specialized company or expert offering seed mixtures for grazing who will know what will suit your particular habitat (wet or dry, low or high altitude, and so on).

In pastures and areas intended for hay production, a mixture of grass and suitable herbs can be sown. Some suitable examples:

- Chicory (*Cichorium intybus*)
- Anise (*Pimpinella anisum*)
- Common fennel (*Foeniculum vulgare*)
- Chamomile (*Matricaria chamomilla*)
- Buckhorn plantain (*Plantago lanceolata*)
- Dill (*Anethum graveolens*)
- Cilantro (*Coriandrum sativum*)
- Salad burnet (*Sanguisorba minor*)
- Great burnet (*Sanguisorba officinalis*)
- Lemon balm (*Melissa officinalis*)
- Pot marigold (*Calendula officinalis*)
- Dandelion (*Taraxacum officinale*)
- Coneflower (*Echinacea purpurea*)

...and many others. Take care to plant herbs, flowers and grasses that are native to your area.

Many of these herbs don't do well if trampled, and may disappear in a smaller field. Others may plant themselves freely without any human intervention.

If you have nettles growing in your fields, you can leave a patch of them. Most horses, unless they have severe mineral deficiencies, won't eat them fresh but they will gladly eat them once they're cut and dried. It's one of the few plants which will survive well even in a paddock, as long as it isn't heavily trampled.

Suitable herbs can have a positive effect on horse's digestion, and also contribute bioactive substances which herbivores can make excellent use of. Giving our horses the opportunity to select freely from a variety of flowers, grasses and herbs to suit their current needs is an ideal way to keep them healthy and happy, given enough space.

Watch out for calcitriol! Work to keep growth of yellow oat grass in check in your pastures—although it isn't particularly palatable, it has a high concentration of calcitriol (vitamin D) and if grazed heavily, can increase the concentration of calcium in the blood, reduce urinary excretion, weaken bones, and cause cartilage calcification.

Common thyme is used as an herb in folk medicine as well as in the kitchen. The leaves can be used for tea, it has a positive effect on digestion and has antibacterial properties.

A nettle corner left intentionally in a basic paddock.

We don't only help our horses by adding different plants and herbs to their fields, but we can also help the environment: many are host plants for insects and butterflies in larval and adult stages, creating a habitat which supports our horses without sacrificing biodiversity.

Drawbacks to Herbs in Pasture

Too many herbs in grass cover reduces the yield of the area by breaking up the continuity of the grasses and increasing the spacing in the sod. Some herbs are equipped with mechanisms like poor taste or nutrition as a defense against insects and herbivores. If hay fields have too many such herbs, the hay can be poorer quality. Due to the fact that they're harder to compress than grass, they can create more air gaps in bales as well.

The composition of hay fields determines its nutritional value. If it contains highly productive grasses and clover, the hay will be dense and nutrient-rich, but it may be too rich for some horses. Regular feeding of rich hay to horses in light or no work can quickly lead to obesity, especially for certain breeds. On the other hand, although an area with energy-poor (low-sugar) grass will bring in smaller yields in hay, it will be safe and suitable for regular feeding.

Hay should be stored in a well-covered area with good air circulation. Bales shouldn't be put directly on the ground, but on pallets that allow air flow underneath. Hay should be very dry to avoid issues—clover hay, for example, if baled while too wet, can become toxic as coumarins (which produce the characteristic "hay" smell when it's cut) are converted to dicoumarol (which acts as a blood thinner) if it becomes moldy. If hay contains sweet clover, it may be a good idea to have it tested for dicoumarol content as visual inspection isn't the best way to determine toxicity.

Frequently Asked Question:

During the winter, when temperatures are below zero, our two horses (who are on open pasture with access to hay) lose weight—especially our young gelding. Over the summer, they're too heavy and we have to cut back their grazing. Over the winter, though, we have to invent ways to keep some weight on him. We know that some weight loss is to be expected during the winter and is good for a horse who goes into the season overweight, but at the end of the winter, he just doesn't look good. Our vet says he's healthy and recommended blanketing, which helped, but only a little. Can he be helped in some other way, apart from blankets and extra feed?

With broodmares, watch out for wild carrot—Queen Anne's lace (*Daucus carota*)—and wild parsley (*Petroselinum crispum*). These two herbs can act as abortifacients. If you have a herd of geldings or mares not destined to breed, rest easy; they aren't toxic otherwise.

If your horse is blanketed, getting regular access to hay and grain, and in good health, but is still losing weight in winter, it's very likely that he's struggling with your winter climate.

If his condition improved after he was blanketed, try to think about a setup where he doesn't have to withstand high winds in open areas. Provide him with as many places to hide from the wind as possible. This, in combination with his snacks and warm blanket, should help him maintain a better condition. I would also suggest buying (or making) different types of hay: "diet" hay for the summer season and energy-dense hay for winter months.

If you have less space to work with, you can provide herbs which aren't available in pasture with an "herb garden." The plants are protected from being over-grazed or ripped up with the help of a grid. You can use an old tub or a large tire filled with soil, or a raised bed made out of stone or brick.

Organic Composition = Condition

The condition of our horses is determined by the nutrients and energy provided to them in *everything* we feed them. This means we shouldn't just focus on the nutritional value of their grain and supplements, but also—and I would say first and foremost—their forage (pasture and hay). Unfortunately, it seems far too common that people give little to no thought to the composition of the roughage their horses are given.

Most horses who graze freely all day, are given hay in their basic paddock at night, and aren't in work (and look to be in good condition) are likely to have their nutritional requirements met without adding concentrates to their diet. If a horse is in light work, this may still be sufficient, but if they start to lose weight, grain or concentrates should be added. For horses in heavier work, additional feed is a must.

Establishing a Pasture

When creating pastures, don't focus only on high-yield or energy-poor grasses; it makes the most sense to sow the area with a balanced mixture intended specifically for horse pasture.

Types of Hay

- **Meadow hay:** A combination of grass species cut from natural pasture. It contains a greater variety of plants.
- **Legume hay:** Typically alfalfa or clover. Higher nutrition than meadow hay (it contains more digestible protein, plenty of calcium and phosphorus, beta-carotene, vitamins D and E, has a variety of minerals and is lower in sugar). Can be given as up to ⅓ of a horse's daily ration (more for lactating mares) after habituation. May be advantageous for growing horses and can help horses prone to gastric ulcers.
- **Grain or cereal hay:** Of these varieties, typically oat hay is best for horses. It should be cut when the seed is out of the milk stage (when the seed head is squeezed, if it oozes a milky sap, it's the milk stage). At this point, the stem will still be palatable. Oat hay is high fiber, low protein, and should be adequate to meet the nutritional needs of horses. It shouldn't be given to insulin resistant horses, as it tends to be higher in nitrates and sugar.

Hay should be stored in a well-ventilated, covered area.

If pasture is sown with energy-poor choices, the hay it produces won't be sufficient even for out-of-work horses. However, if the growth established is too rich for the intended horses (for example, horses not in heavy work), we run the risk of health issues and obesity.

Think about the composition of your pasture (and hay) as the main building block for your horses' health and condition. A horse's condition is more effectively influenced by changing the nutritional value of the roughage than it is by changing work or concentrates.

If your horses are gaining too much weight despite eating only hay, there are a few possible solutions. Straw can be added to the hay to "water it down", hay nets can be switched out for ones with smaller holes, and exercise can be increased. If these measures don't help, the problem is very likely the hay and pasture grass. If your horses are eating unnecessarily high-nutrient forage for their condition, sometimes the best option is to find a supplier who has lower-quality hay. If you make your own hay, wait a bit longer to harvest at the beginning of the season. Another option is to leave the grass after mowing to get rained on once or twice (being sure it's perfectly dry before baling, of course) to reduce its nutritional value. Lastly, add seed for low-sugar grasses when it's time to re-seed fields that will be used for the rounder members of the herd.

If your horses are too thin even with free choice hay, it's often easily fixed by providing better quality hay, or by substituting alfalfa for a portion of it. When it's time to re-seed fields, it can help to add a mixture of higher-yield grasses (like alfalfa) to make up 10–20 percent of the total grass available. It's important to mention that fresh alfalfa is far richer, and should make up a much smaller amount of a horse's diet, than when it's dried.

Some people feel alfalfa is too rich and refuse to add it to their horses' feed. I see no reason for this—I've been incorporating it, green and dried, into feed rations for the past thirty years. Alfalfa has a deep taproot that tolerates dry conditions. This allows for enough hay even during drought. When other grasses go dormant to survive the summer heat, the only green in the fields are the patches of alfalfa.

It's less able to withstand trampling than other grasses, so in pastures where horses are left out long-term, it won't hold up to the stress. In areas where horses are only allowed for short times, however, it's a great option.

The maximum amount of alfalfa sown into any field meant for hay is 30 percent—and some areas shouldn't have any, to ensure there's always some hay available that doesn't have any alfalfa mixed in. In pastures where horses might be left indefinitely, no more than 10 percent of the area should be sown with alfalfa.

The deep root network of this plant enriches the soil profile where it's planted, adding a significant amount of organic matter. In times where there is a scarcity of organic fertilizer, alfalfa can be a significant factor in increasing soil fertility.

It also has a high nutritional value, being one of the

The protein, mineral and vitamin content in hay depends primarily on how leafy it is. Hay that is cut low will be more nutritious (fewer stems). Because of this, hay should be turned early in the morning, while it's still damp. If it's turned later in the day when the heat has dried the hay and made it crunchy, leaves will break off, leaving only the stems to gather up and bale. Especially with alfalfa, it's important to keep this in mind if we don't want to leave nutrition behind in the field...

forages with the highest crude protein content while at the same time being easily digestible. In this sense, it should be treated the same as grain: some horses may need it, others may not.

It should always be fed in measured amounts; unlike meadow hay, which horses can have access to 24/7, alfalfa should never be fed in unlimited quantities.

For me personally, during years where there's a lot of alfalfa hay, I like to feed it instead of grain, which I completely leave out. With lactating mares, however, I decide based on their condition. I'm comfortable giving a lactating mare half meadow and half (possibly a bit more) alfalfa hay. For horses in heavier work (not high-performance horses), I'll give more alfalfa. For high-performance horses, nutritionists no longer like to see alfalfa hay, as it adds a lot of protein and horses use a lot of energy (needed for peak performance) to process protein, so this can unnecessarily overload the liver's capacity.

Palatability is related to the type of forage and its digestibility. The amount of lignification (hardening of cell walls) has a greater influence on nutritional value than any other factor. In older plants, the majority of cells will have hardened, making the plant less digestible. This can be an advantage for obese horses—this hay will fill them up without being too calorie-rich.

In addition, they excrete excess nitrogen in their urine (in the form of ammonia), which ruins the air quality in a stable (these horses aren't usually kept outside).

If you're not sure what to do, consult a nutritionist. Be sure to give them an accurate account of your horse's exercise routine.

Buttercup is poisonous when alive, but loses its toxicity when cut and dried.

Recommended minimum grazing heights:

2"–3" minimum (fine blades, short species)	3"–5" minimum (taller, high-yield species)
Bluegrass	Orchard grass
Bermudagrass	Timothy
Perennial ryegrass	Tall fescue

For more information, see the equine pasture management page on the PennState website: https://extension.psu.edu/basic-pasture-management-for-the-equine-owner

Grass species and their ability to withstand trampling:

more durable	more delicate
Kentucky bluegrass	Prairie grass
Tall fescue	Timothy
Clover	

(The remaining species are somewhere in the middle of the spectrum, where a lot depends on the specific conditions of the location.)

Watch out for red sorrel!

Rumex acetosella, also known as sheep's sorrel, field sorrel, and sour weed, is native to Europe but has been introduced to the US, and can move into bare spots in pastures. Horses will only graze on it if nothing else is available due to its bitter taste, but if enough of it is eaten, it can be highly toxic—it contains oxalic acid, which, in large enough doses, can affect horses' ability to process calcium and cause kidney failure (it does the same thing in humans). Red sorrel has a deep taproot which can go several feet deep—it should be pulled only when the soil is very wet to get as much of the root as possible.

Red sorrel's pollen can bother people with allergies, too. Its only benefit is as a host plant for caterpillars of some species of butterfly—this is the only possible reason to allow it to survive in a small corner of the pasture.

Frequent mowing can weaken this tenacious weed and eventually kill it, but two to three times per season is nowhere near enough to get the job done.

Use of Meadows and Pastures

If you have enough space, the ideal way to manage fields is to alternate them so that at least once each year each area is left alone until it's mowed (for hay or haylage)—which is to say, give it one growth cycle without allowing any grazing. This way, the sod will be able to thicken, and shorter grasses which are suitable for horses will be allowed to expand into more areas.

Frequently Asked Question:

Can a horse survive only on pasture?

It depends on the individual horse and the composition of the grazing area. A horse's digestion is very specific: *enzymatic digestion* occurs in the stomach and small intestine (foregut), and *microbial digestion* takes place in the cecum and colon (hindgut). Proteins, fats, simple sugars and starches are digested enzymatically in the foregut—part in the stomach, and most in the small intestine. If a horse eats more than 3–4½ pounds of grain or concentrates in one meal, he runs the risk of having excess sugar in the hindgut. The bacteria in the hindgut require a higher pH to function; the lactic acid produced by the fermentation of excess sugars lowers the pH, and the bacteria needed to digest fiber die off. This acidic environment in the intestines can have serious health consequences.

A horse is evolutionarily adapted to grazing on dry, poor grass which is less energy-rich. The ancient horse needed to be able to extract the maximum from a minimum. If our horses weren't expected to be athletes for us, and we focused less on high performance when breeding, a low-nutrient, extensive pasture would be enough to meet the needs of most horses.

It's only when we increase their workload that there's a real need for higher-energy feeds.

Despite this fact, it's common practice to feed horses some kind of grain or concentrate twice

Horses aren't made to consume large amounts of simple sugars and starches. They're excellent at extracting enough energy for a day's fiber intake. For overweight or out-of-work horses, the addition of concentrates is unnecessary and possibly even counter-productive.

each day, with these same horses spending the rest of the day standing over nutritious bales of hay. This is a far cry from traveling far and wide in search of sustenance. Most horses nowadays take in quite a bit more than they need, which can quickly lead to obesity.

The best way to avoid long visits with veterinarians and nutritionists to solve our horse's health problems is to focus on the best pasture for the given location and climate. You can plant a great mix of grasses meant especially for horses, but you'll only be successful if the conditions where it's sown are suitable. There's no way to make a moisture-loving plant tolerate drought, and vice versa.

Basic enclosures that can be opened or closed off at any time make it easy to regulate how long horses stay on pasture without restricting their time outdoors.

If there are certain horses with a tendency towards weight-gain, the amount of time spent on pasture should be limited. Otherwise, their grass intake can be limited by using a grazing muzzle.

All owners are responsible for their own horse, even if they're boarded on someone else's property. They should know the risks to their horses, or any limitations regarding pasture turnout. These should be communicated to anyone responsible for the horse's care so that an informed decision can be made as to where and with what group the horse should be placed. If you keep your horses at home, you must take the necessary steps yourself to minimize any risks from overgrazing.

An obese horse is an unhealthy horse. A grazing muzzle can often help limit grass intake while allowing the horse to stay with the group. Although some horses will fight wearing it or learn how to take it off, some get used to it without any problems. There are newer kinds than the basket design pictured here which many horses find far more comfortable. You may need to experiment to find the right fit for your horse.

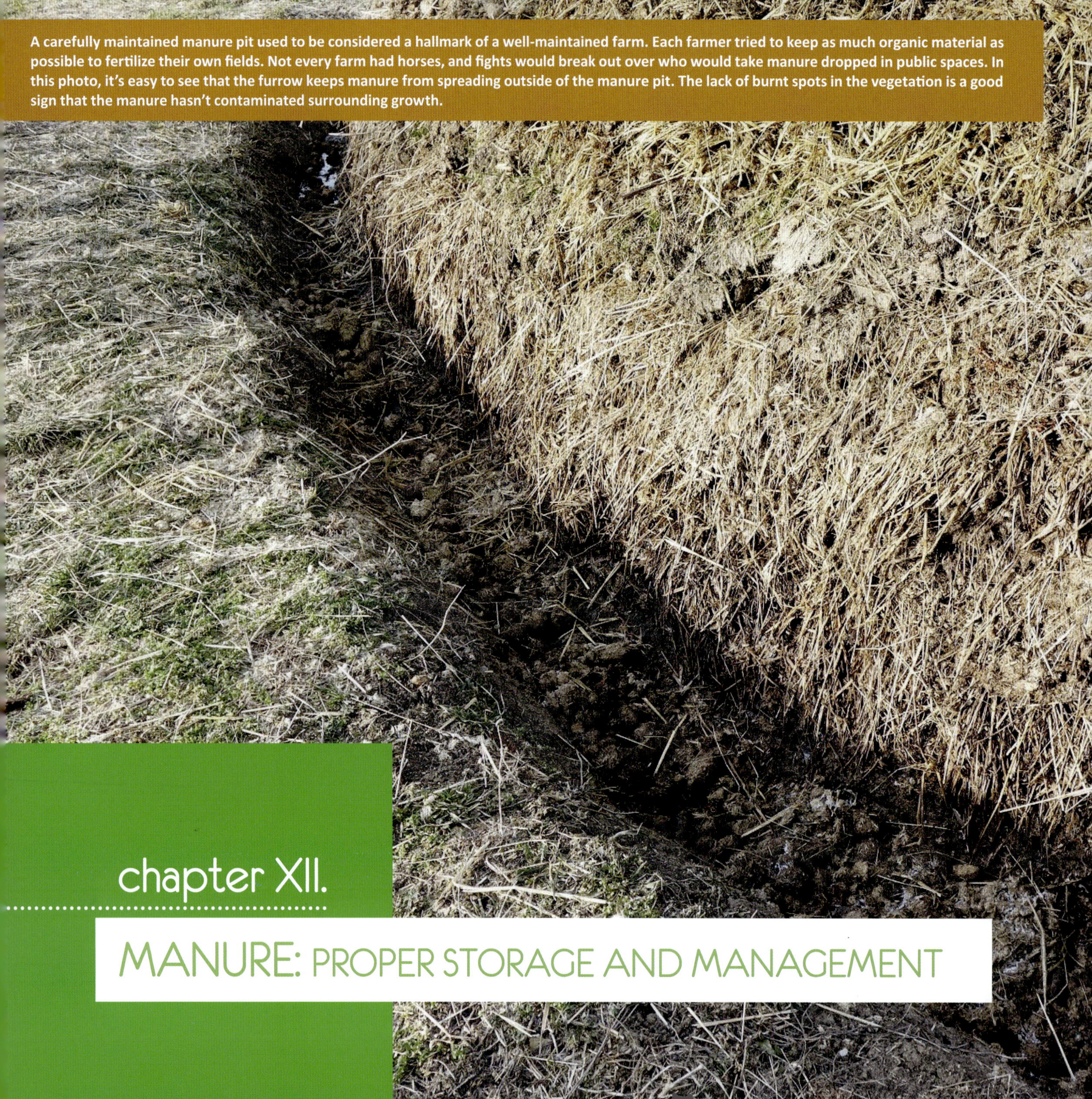

A carefully maintained manure pit used to be considered a hallmark of a well-maintained farm. Each farmer tried to keep as much organic material as possible to fertilize their own fields. Not every farm had horses, and fights would break out over who would take manure dropped in public spaces. In this photo, it's easy to see that the furrow keeps manure from spreading outside of the manure pit. The lack of burnt spots in the vegetation is a good sign that the manure hasn't contaminated surrounding growth.

chapter XII.

MANURE: PROPER STORAGE AND MANAGEMENT

Manure

Manure collected from paddocks, pastures or stalls needs somewhere to go—many farmers don't think much about it, leaving it in a huge pile for years before taking it back to the fields. Others have it moved several times each week.

But where to put the poop, so that it can become the compost that will fertilize the fields? A properly constructed structure is key.

The only way to avoid the need to create a location for storing manure is by immediately spreading it in fields or sending it off to other farmers for their use. For everyone who can't do this, it's important to think carefully of how best to build in an area near either horses or homes and, most importantly, to follow any applicable laws in your area.

Storage

Some agricultural facilities simply store manure in an unused livestock lot. These piles of manure can be risky for the surrounding environment if sufficient precautions aren't taken.

Solid and liquid waste from horses, mixed with bedding material (either straw, sawdust, or shavings) will, thanks to fermentation during proper storage, turn into compost. Compost is a wonderful fertilizer, which is invaluable for healthy pastures.

Solid organic fertilizers and farmyard manure should not be stored long-term on farmland before being used, especially during warm weather. Be sure to know the rules in your area for maximum allowable time and suitable storage options.

After the manure is removed, the soil in the area will have elevated levels of nitrogen and other nutrients. Less nitrogen is left in soil (up to a depth of approximately 4 feet) by horse manure (as well as goats and sheep) than by poultry or cattle.

Agricultural universities have studied various sites, where they monitored the pollution surrounding these manure lots and the contamination of subsurface waters and topsoil layers. They concluded that lower layers were less contaminated than expected. Increased amounts of nutrients in the soil mostly occurred to a depth of about three feet directly below the lot, with amounts decreasing rapidly as the distance increased.

Rules for the On-Property Storage of Manure

- Observe a minimum recommended distance of 100–200 feet between water sources (such as wells, wetlands, streams, ponds or lakes) or erosion-prone areas and the storage site.
- Prevent runoff and leaching by using a well-drained, slightly sloped area. Use a well-vegetated filter strip or a dug furrow to catch

Vzorně vrstvené a udržované hnojiště.

runoff or add enough soil or straw to the manure to absorb liquid.

- The storage site must not pose a danger to the environment. The site should be compacted to prevent leaching of nutrients into the soil, and the pile should be kept as dry as possible.
- Nothing should be added to a manure pile that would disrupt the development of agricultural crops or disrupt the food chain.

Manure and Nutrient Management Plans

You don't need to be a specialist to create a manure management plan, and depending on where your farm is located, you may need one. Elements in these plans usually include: farm features, methods for spreading and/or storing manure, pasture management, and where animals are concentrated.

In the US, some states' environmental regulations require a manure management plan. Although the number of animals being kept may determine which plan you need, there are areas in which keeping just one or two animals is enough to require you to develop a written plan. Make sure you know what regulations apply where you live.

A nutrient management plan, on the other hand, needs to be written by a specialist and submitted for approval. These plans may be worth the difficulty, as they can offer limited liability in the event of a civil complaint. (This information comes directly from PennState's page on manure management plans, as of the time of writing: https://extension.psu.edu/have-a-horse-or-steer-in-your-backyard-you-need-a-manure-management-plan. You may need to keep different considerations in mind, depending on where you are operating.)

Winter Management and Storage

Areas that collect manure during the winter (which offer protection from bad weather when horses can't be on pasture) are potential sources of water pollution because the horses are concentrated in a smaller area. Prevention of leaching of harmful materials from this area should be part of your manure management plan.

The manure from areas used during the winter should be cleared and responsibly stored or spread at the end of the season.

If these areas aren't cleared once they're no longer in use, they can become a breeding ground for parasites.

Avoid turning your basic paddock into just a place for manure storage by collecting manure and changing bedding regularly.

A manure bin is an above-ground structure used for storage. It can be open or covered. The bottom should be impermeable.

chapter XIII.

WINDBREAKS IN PASTURES

A makeshift windbreak made of old boards.

The time put into building windbreaks now will serve us well, and years later, will certainly be better for the land and for future generations than the mindless eradication of any vegetation which can't be grazed. Forget the notion that a pasture is a bare area which only serves for grass yield and nothing else. Take charge of the land you have—you can pamper your horses with a varied landscape. Start planting as soon as possible!

You wouldn't want to plant a tree windbreak in a raised bed, but (if it's on your property) a raised bed can be created in front of a windbreak.

A windbreak will generally protect an area about ten to fifteen times as large as the windbreak is high. This doesn't mean there will be absolutely no wind—that's not how a windbreak is designed to work. A quality windbreak will moderate the wind, not completely blocking it. If, for example, you build a stone wall in the wind's path, you'll create a bit of protection just behind the wall, but the wall will create eddies (whirlwinds) that will create more turbulent wind conditions.

Winter Windbreaks

To protect property from winter winds and snow drifts, windbreaks should be built along northern and western edges of the area (winter winds predominantly come from the north and northwest).

Summer Windbreaks

If you're concerned with lessening drying summer winds, windbreaks should be placed along southern and western edges (as wind switches from north to south in the summer months).

Types of Windbreaks

Impermeable

This type of windbreak is created by densely planted rows of shrubs and trees, so that they form an interconnected wall. It should be mostly trees, with shrubs in a row in front (they can be planted behind the trees as well).

Semi-permeable

Like the windbreak above, these also have several rows of trees and a shrub layer, but the tree crowns aren't as inter-connected, and the shrub floor is a bit less dense. This way, the windbreak can function well without creating air eddies.

Permeable

These are created with a row or two of trees only, without a shrub layer. This kind of windbreak is becoming less popular, as they can create a "wind tunnel" (where the wind gains strength as it passes through the narrow openings made by the trunks).

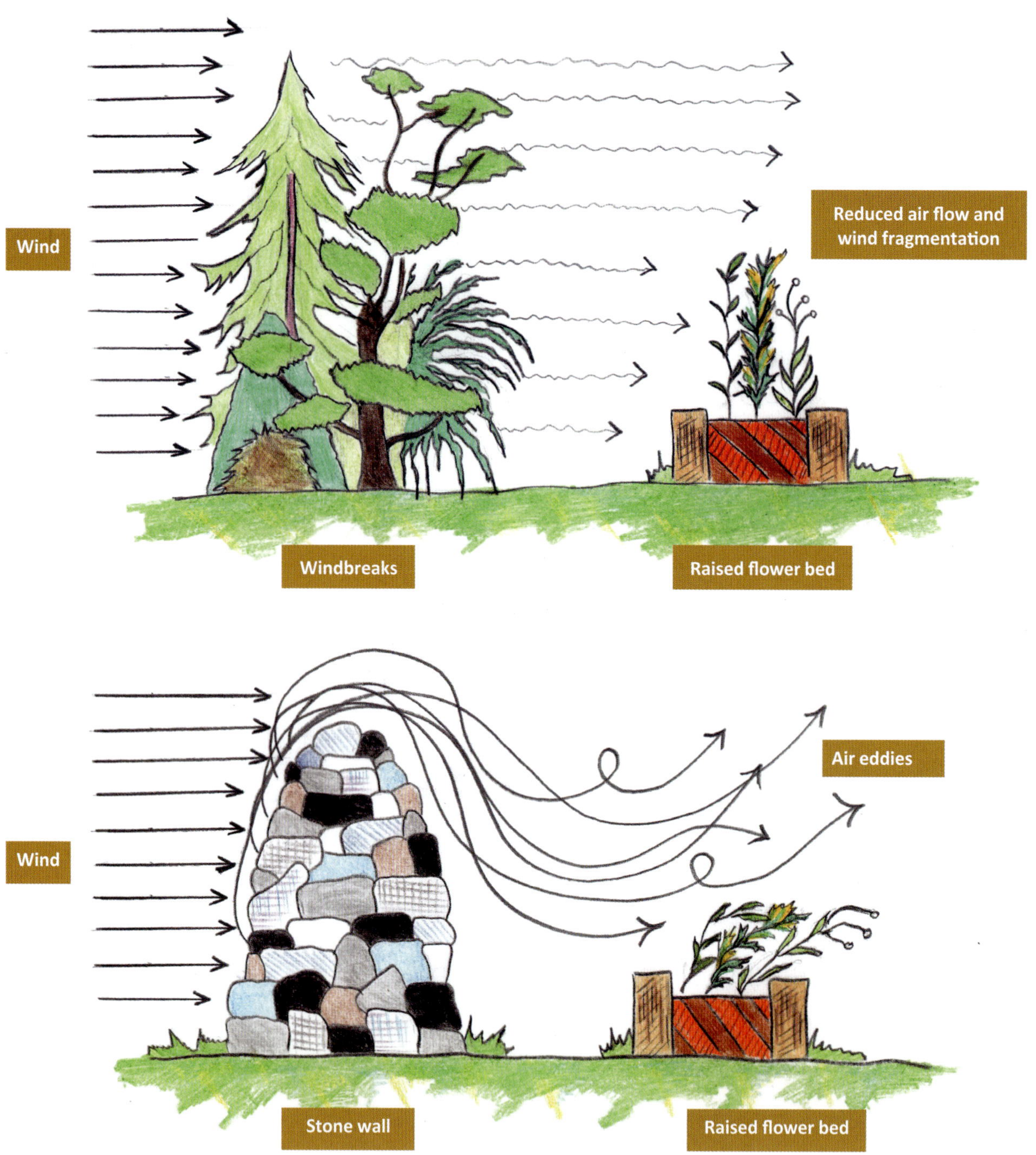
Wind
Reduced air flow and wind fragmentation
Windbreaks
Raised flower bed
Air eddies
Wind
Stone wall
Raised flower bed

A windbreak isn't intended to stop the wind completely, but to break it up so it's more moderate. A few years after planting, this windbreak's growth will be much denser and will keep horses considerably more comfortable.

Suitable Trees

Pine, oak, maple, elm, American basswood, red alder and black walnut trees all have deep roots which anchor them well. Trees like birch, ash and poplar have shallower roots and should be planted in between belts of stronger trees for protection.

Suitable Shrubs

Shrubs protect the roots of your windbreak trees from the drying effects of low air currents. They also help to trap snow, dirt and weed seeds so less is blown onto the pasture. The following chapter goes into more detail about the best types of shrubs for a windbreak.

Planting Your Windbreak

Trees should be planted in spring or autumn, in holes approximately one foot deep (depending on the age of the saplings). Rows should be planted at least five to seven feet apart (or wide enough for whatever machinery you'll need to get through for maintenance work). Spacing between trees within the row should be about five feet apart. Shrubs should be planted according to their size at maturity: when fully grown, they should be partially intertwined to form a living fence. Consult an expert about optimal spacing for the particular shrubs you choose.

Be sure to support your saplings: place stakes on the south side to protect them from the fierce summer sun. Always hammer the stake into the ground prior to planting to avoid damaging the sapling's roots.

If your windbreak is close enough to your basic paddock or pastures for leaves to drop or blow into the enclosures, be very cautious about the tree and shrub species you choose. For example, black walnut and red maple leaves are very toxic to horses. Know the varieties on your property and assess their safety.

A multi-level windbreak protects this part of the pasture effectively from winds that would otherwise blast through from open space on the other side. In this case, it also provides good protection against spraying on the neighbor's fields.

Windbreak Care

For the first three to four years, a newly planted windbreak needs extra care. The rows should be watered and weeded regularly (in place of weeding, you can add mulch or plant grass which can be mowed to keep weeds at bay). Trees should be fertilized according to their specific needs.

Windbreaks as a Useful Part of the Landscape

Strong winds aren't fun for humans or horses, and they aren't good for plants either. A natural windbreak not only adds beauty to the landscape, but also food and shelter for a variety of wildlife. Insects and birds will move right in, and shrub understories offer crucial protection for small animals from predators. If you add stones under the plantings, you'll also create a safe spot for lizards, toads, and snakes.

Fallen leaves, flowers and fruits improve the soil microflora. Since the aim is to plant native species for your region, your windbreak can be a source of seeds for the surrounding area—birds will deliver them for free.

Hard Pruning (Coppicing) and Windbreaks

A natural windbreak benefits our horses by taming the wind and offering shade. It can also benefit us by providing wood for heat.

Whatever kind of windbreak you plant, it shouldn't have gaps (like a path leading through). A gap defeats the purpose of the windbreak, allowing for wind to significantly increase as it passes through.

It's not impossible to grow your own fuel for wood-sourced heat: there are plenty of undemanding, fast-growing trees. Some trees will naturally overgrow once they're cut down; by coppicing trees (cutting them close to the base to prompt multiple new trunk shoots to grow from the stump) in a years-long cycle, you can rejuvenate some trees rather than eliminate them. Each species has its own "youth cycle." As trees age, they don't grow back as well. Make sure your tree is established enough to tolerate hard pruning, but stop cutting back before the tree is too old to recover.

Growing species which can be coppiced can be easy around pastures, meaning an easy source of heating wood for winter. If you plant in several rows, gradual cutting back won't affect your windbreak's utility. The best time to do this is when the trees are dormant (usually winter). Caring for forest trees isn't hard—it's actually much easier than worrying about the demands of ornamental trees—as long as you choose species that grow well (and look good) when cut back. Many also have value for wildlife as food or shelter. The table on this page lists some of the trees and shrubs with coppice potential. Some are more tolerant of pruning than others; be sure to research the particular growing conditions and needs for your chosen species. Trees which can often be suited to hard pruning will have minimal bleeding, are deciduous, shrub-like, suckering and non-flowering (although there are exceptions). If the trees are in an area where they may be browsed by deer, cattle (or horses!) cutting higher (aka pollarding) can be a simple solution to protect new growth.

Trees and Shrubs with Coppicing Potential
Alternate-leaf dogwood (*Cornus alternifolia*)
American beech (*Fagus grandifolia*)
American chestnut (*Castanea dentata*)
American elm (*Ulmus americana*)
American hazelnut (*Corylus americana*)
American hornbeam (*Carpinus caroliniana*)
American sycamore (*Platanus occidentalis*)
American witch hazel (*Hamamelis virginiana*)
Basswood (*Tilia americana*)
Black willow (*Salix nigra*)
Carolina ash (*Fraxinus caroliniana*)
Eastern red cedar (*Juniperus virginiana*)
Green ash (*Fraxinus pennsylvanica*)
Oak (*Quercus*)—all native species
Red elm (*Ulmus rubra*)
Roughleaf dogwood (*Cornus drummondii*)
Shortleaf pine (*Pinus echinate*)
Virginia pine (*Pinus virginiana*)
Winged elm (*Ulmus alata*)

Many varieties of willow are suitable for coppicing, and can be cut even when used as a hedge for pastures. Check to see whether the kinds of willow native to your area are suited for it. If your willow stakes take root and grow well, don't be afraid to cut them!

The Japanese poplar is a non-native tree that is quite popular due to its fast growth rate and the heat value of its wood. It's non-toxic to horses and can work well as part of a windbreak.

For your "horse keeping" windbreak, choose trees and shrubs which are suitable for your region's soil and climate and which will thrive in your particular location. Think about how well they may survive a hard winter or a dry summer, for example. Also, plant deliberately—don't support the propagation of invasive species!

Depending on the type of root growth, tree species, and local conditions, saplings can be ready for coppicing after 3–7 years. This type of pruning can be performed over and over again without killing the tree—if done correctly it can extend the life of the tree.

The stump will regrow faster than a newly planted sapling can grow, since it's growing from a fully preserved and established root network. This network continues to control erosion, and the careful cutting of selected trees lets more sunlight into the understory, allowing shorter vegetation to thrive.

An entire forest can be coppiced, if the trees and woody plants can regenerate from buds growing on the roots and stump. This ability to form root or stump offshoots can vary from species to species. This ability also decreases with age, the strongest offshoots forming from young trees. Growth slows down after about 40 years. Oak and hornbeam have a long-lasting youth cycle. Some trees with a shorter cycle are ash, maple, and birch.

Soil can become exhausted and depleted of nutrients after about three cycles of coppicing, leading to slower and less abundant growth. Fertilization (with composted horse manure) can help maintain a fertile habitat for effective and productive coppicing.

Although we may not be able to engage in this practice on an entire forest like our predecessors, your piece of forest-windbreak can contribute to a happier life for your horses and to a better natural habitat in your community. Besides, what can be more wonderful than wrapping up a hard winter's day of work by sitting by a warm fire?

chapter XIV.

HEDGES FOR HORSES

Why plant hedges? Apart from aesthetics, by adding and caring for hedges, we can make a small contribution to protecting nature. While some neighbors may pave much of their yards, and others transform theirs into a uniform lawn resembling a golf course, you can take the opportunity to create an ecologically diverse landscape. It's absolutely possible to have a nice dry paddock and still have room to grow a variety of plants. With a little effort, horse keeping can go hand-in-hand with environmental stewardship.

Horses are happier if, in addition to pasture, they're given higher shrubs, wooded areas, bodies of water, and planting strips (hedges that the horses can walk around). This can also have a positive impact on nature by providing shelter for birds, small mammals, and a host of insect populations. If you want to help your neighborhood and the ecosystem, the answer is easy: just plant. And plant. Plant thoughtfully, responsibly, diversely.

Trees should be planted at least three to four feet from property lines. Consider the mature size of the tree when deciding where to plant.

After just one year, these juneberry plants have already grown into the beginning of a lush, compact hedge. They can be mulched with composted horse manure and can also tolerate fresh manure thrown on them from time to time.

Why plant hedges?

- They have aesthetic value.
- They function as a biocorridor.
- They can delineate rings, paddocks, and pastures.
- They form a buffer between horses and busy roads.
- They can shield horses from passers-by (making it difficult for strangers to pet or feed them).
- They act as a retaining wall (containing dust and spray from neighboring fields and gardens).
- They act as a barrier to the seeds of weeds and poisonous plants.
- They reduce noise.

The one golden rule when planting hedges around horses: never plant anything poisonous near farm animals, even if you really like it! It's understandable that you don't want to reach for the saw if your new property comes with a beautiful rhododendron, for example, but you should take the necessary precautions to keep leaves from blowing into areas that the horses have access to. Even if the given plant is one horses don't usually bother with, it only takes one curious or bored horse to have a problem on your hands.

Since it's not necessary to start with large plants when growing a hedge, you can allow yourself to think about exactly what you would like on your land without worrying as much about cost.

What *Not* to Plant Around Horses

- Cherry laurel (*Laurocerasus officinalis*)
- European spindle tree (*Euonymus europaeus*)
- Common ivy (*Hedera helix*)
- Holly (*Ilex* spp.)
- False cypress (*Chamaecyparis* spp.)
- Barberry (*Berberis* spp.)
- Firethorn (*Pyracantha* spp.)
- Savin juniper (*Juniperus sabina*)
- Scotch broom (*Cytisus scoparius*)

- Maple (depending on species, can be dangerous seasonally or year-round)
- Horse chestnut (*Aesculus* spp.)
- Dyer's broom (*Genista tinctoria*)
- Alder buckthorn (*Frangula alnus*)
- February daphne (*Daphne mezereum*)
- Rhododendron (spp.)
- Common privet (*Ligustrum vulgare*)
- Smoke tree (*Cotinus coggygria*)
- Buckthorn (*Rhamnus cathartica*)
- European bird cherry (*Prunus padus*)
- Golden chain tree (*Laburnum anagyroides*)
- Yew (*Taxus* spp.)
- Arborvitae (*Thuja* spp.)
- Wisteria (spp.)
- Invasive honeysuckle species (*Lonicera* spp.)—not to be confused with native bush honeysuckle (*Diervilla lonicera*)
- Boxwood (*Buxus sempervirens*)
- Bladder senna (*Colutea arborescens*)

The abbreviation "spp" indicates that within this genus there are several subspecies, and none of them are suitable for planting where horses may get at them.

Lilac hedge surrounding paddocks: although lilac contains the irritant syringin, which can cause dermatitis in people, it isn't poisonous and doesn't seem to affect horses.

An English farm with paths surrounded by hedges.

So, in the search for fast-growing shrubs which can form a functional hedge soon after planting, what's left after invasive and toxic plants are ruled out? Some suitable recommendations:

Deciduous Shrubs and Trees

American Hornbeam **(Carpinus caroliniana*) and* *Scarlet Hawthorn*** **(Crataegus coccinea*)***

These slower-growing species, with a little patience, can create a lovely and functional hedge (which is safe for animals) in a few years' time.

Hornbeam is a hardy and adaptable plant which tolerates a variety of conditions once established. It can withstand brief periods of drought and overwatering. It can grow in shade or sun, having a denser shape grown in the sun. Hornbeam tolerates pruning fairly well. Although hornbeam is deciduous, it holds its dried leaves until spring when new leaves sprout, providing a safe winter habitat for wildlife. It's used as a food source by small mammals and birds and is an important host plant for several species of butterflies and moths. Hornbeam is resistant to diseases and pests.

(References: https://wp.towson.edu/glenarboretum/home/american-hornbean/

https://piedmontmastergardeners.org/article/consider-a-hornbeam/)

Hawthorn is considered a medicinal plant, so we don't have to worry about it near our farm animals. It's prone to diseases that can be transmitted to apple trees; a good rule of thumb is to avoid planting hawthorn near fruit trees. Being impenetrable due to its thorns, this plant is ideal for marking a pasture's edge.

Scarlet hawthorn (*Crataegus coccinea*), an eastern US native, is an undemanding plant which is easy to care for. It can adapt to a variety of climates and conditions but prefers full sun. Its flowers attract pollinators and its berries are a valuable food source for birds and other small mammals. The thorny branches offer shelter and protection for wildlife. An added benefit of the hawthorn—it tolerates pruning very well and can even withstand being cut down to the ground, making it an excellent choice for a hedge.

*Golden Currant (*Ribes aureum*)*

Golden currant, although native to the Rocky Mountains, has escaped cultivation and become established in eastern states as well. This is an adaptable shrub which can tolerate standing water and drought, and can grow in full sun or shade. The flowers are a nectar source for hummingbirds, butterflies and bees, and the fruit is eaten by birds. Its shape and moderate growth rate make it a good choice for a hedge plant.

*Juneberry/Serviceberry/Shadbush (*Amelanchier lamarckii*)*

This lovely shrub or small tree can grow from 12-25 feet tall. The foliage offers vibrant fall colors, the flowers which bloom in the spring are attractive to pollinators and are followed by fruit which can be made into jam or left for the birds. It serves well as a hedge or screen—if left to grow taller, it can form a boundary that the horses can see through to satisfy their need to monitor their surroundings. Serviceberry is best kept out of reach of the horses for the sake of the plant's health and appearance, as they may find the berries to be a tasty treat.

The Siberian elm hedge separating these plots of land from the neighboring buildings is 4 years old—and located in Europe, where these trees are native. They're invasive in the US, however, and should be removed if you find some on your property.

Silky Dogwood (*Cornus amomum*)

This shrub thrives in partial to full shade and would love to grow on a stream bank or in a flood plain as it requires plenty of moisture. It grows up to 15 feet tall and has showy white flowers. It attracts both birds and predatory insects (insects which prey on pest insects). All dogwoods are considered non-toxic to animals, so are safe to grow around horses. (Reference: https://www.wildflower.org/plants/result.php?id_plant=coam2)

Red-Twig Dogwood (*Cornus sericea*)

This native dogwood shrub likes moist, well-drained soil but is fairly adaptable to a variety of conditions. It spreads by suckering and grows to approximately 12 feet tall, making it a plant ready to spread into a hedge. Its summer leaves and flowers are followed by a lovely winter show of red branches. This plant is beneficial to a variety of wildlife. (Reference: https://www.wildflower.org/plants/result.php?id_plant=cose16)

Roses

If rose plants are allowed to intertwine as they grow, they'll create a virtually impregnable fence line due to their thorns. Most homeowners grow roses up against a fence to support the plant. Once the plants develop, they'll hide the man-made fence almost entirely. Consider planting one of the 20 US native rose species—they require less maintenance than their exotic counterparts and are far more beneficial for our pollinators.

Highbush Blueberry (*Vaccinium corymbosum*)

A slow-growing deciduous shrub, blueberry thrives in poorly drained areas. This plant requires acidic soil; consider grouping with other acid-loving plants. Given the right growing conditions, blueberries are low maintenance. It will eventually grow from 6 up to 12 feet high. Choose different cultivars to extend the time you'll have berries. Although blueberries are safe for horses to eat, don't allow them unrestricted access to the fruits—they should only be eaten in small quantities. (Reference: https://plants.ces.ncsu.edu/plants/vaccinium-corymbosum/)

Evergreen Hedges

Eastern Arborvitae (*Thuja occidentalis*)

This tree is also called whitecedar (one word, to differentiate it from the family of true cedars), and is a medium-sized tree that tends to grow to an average height of around 50 feet. They do tend to be attractive to deer, and will struggle in difficult growing conditions. In nature, they are native to Canada and the northeastern US, and are well-suited for wet, cool conditions. Their branches are capable of taking root if the main tree falls. There are multiple ornamental varieties specifically cultivated for use in hedges as natural fencing.

Wax Myrtle (*Myrica cerifera*)

Also called southern wax myrtle, southern bayberry, candleberry, bayberry tree, and tallow shrub, this species is native to the southeastern US, and can grow successfully up to around the latitude of New York City. It is capable of weathering coastal storms, long droughts, and bouts of extremely high temperatures, and has uses in candle-making (as some of the common names suggest) and as a medicinal plant, in addition to its ornamental appeal.

***Red and Blue Spruce* (Picea) *and Balsam Fir* (Abies)**
These non-toxic conifers can create a very tall fence line which looks and smells beautiful year-round. Unfortunately, they grow very slowly, and pruning and maintenance aren't simple. These plants will *not* recover from a hard prune. In older trees, the lower layers of branches may die off. If you choose to grow a spruce fence, be sure to consult with an experienced arborist.

***Native Cane* (Arundinaria) *Instead of Bamboo* (Bambusa)**
Bamboo may seem like an attractive option, considering how quickly it grows, but in the US, it's a highly invasive plant, and some states have passed laws regarding the planting of bamboo on one's property.

Instead of the non-native *Bambusa*, consider planting *Arundinaria*, which is a US-native cane. It's an adaptable plant capable of growing up to 30 feet tall; it will also create a beautiful living fence, and has value as a forage plant for horses. (Reference: https://www.fs.usda.gov/database/feis/plants/graminoid/arugig/all.html)

Rose bushes as a hedge—if they grow well and intertwine with each other, roses can create an impregnable fence, guarding their space with sharp thorns. Roses will need a man-made fence for support, but the plants will eventually hide the fence almost entirely.

Buying expensive seedlings is a good investment, but only if we care for that investment. If you're going to plant in the wrong location and fail to care for the young plant until it's established... you may as well spend your money on something else!

chapter XV.

NATURAL FENCING

In the previous chapter, I discussed some of the possibilities for using hedges as natural boundaries. However, a natural-looking fence can be created without having to plant hedges if there's a need (for example, the land is leased) for a temporary solution: we can stack or weave twigs and branches to create fencing.

Older fences in the countryside are sometimes woven (vertically and horizontally) out of natural wicker or thicker branches. Later on, planks which were unsuitable for building were densely grouped and pounded vertically into the ground. Other materials widely used for fencing: damaged or chipped boards, logs and split rails. Where landowners had more money, planks were hammered onto posts. A fencing system I'd like to discuss here is one not seen very often in the US: the Benjes hedge.

The Benjes Hedge

The Benjes hedge, or *Benjesheck*, has some popularity in Europe, but isn't very well known in the United States. It first appeared in the 1970s, when Hermann Benjes, an ecological horticulturist from Lower Saxony in Germany, decided to use branches—and sometimes even entire shrubs—that had been cut during regeneration pruning to create boundaries in agricultural fields.

The resulting "dead-living fence" made out of the deadwood left from pruning is a simple way to close the loop and effectively recycle deadwood to create much-needed solutions not only for your challenges as a horsekeeper, but for ecological issues as well.

The fence's shape is defined and maintained by stakes hammered into the ground (an option for greater stability: weave long flexible branches through them before adding your cuttings).

These hedges can be an important facet of permaculture by using waste material instead of adding it to landfill. This wood, as it goes through varying stages of decomposition, will be a space for fungi, mosses and microorganisms that live on and in the wood, and a place for food, shelter and overwintering for insects (did you know that many insects that depend on deadwood are threatened?), birds and small mammals. Depending on your location, your fence may also host amphibians and reptiles. The increased biodiversity your hedge provides will help keep pests under control without harmful pesticides-creating a naturally balanced ecosystem. Let's not forget the longer-term benefits of the Benjes fence: as wood fully decays, the soil is enriched and becomes a fantastic location for adding new plantings, whether or not you'll move or remove the hedge in the future.

Once a Benjes hedge is established, it can be transformed into a "living" fence with the addition of tall or climbing plants, whether by deliberate planting, or from birds who take shelter in the fence leaving behind seeds as they "fertilize" the space (nature will happily green the space for you, given enough time). It's up to you whether you decide to let nature have free rein or if you would rather be more deliberate in choosing the plants. Be sure to use native plants which benefit your region's pollinators.

If you decide to put up a Benjes hedge, think of the butterflies! After bees, they're some of our most important pollinators.

Creating suitable conditions for their survival is almost necessary these days. Many butterfly species need sunny habitats like meadow or scrub forest, which are unfortunately considered "messy" by most standards. How many of today's manicured yards and gardens can meet these standards? We've carefully removed all the "weeds" and disposed of leaves and branches... look around and see for yourself!

Older, drier Benjes hedges can pose a fire hazard. Be careful in your choice of location so that you aren't increasing the risk of fire during hot, dry weather. If you have the option to do so, it can be a good idea to wet down the fence during extremely dry weather. (It can also pay off to plant horse-safe climbing annuals on your Benjes hedge for quick, eye-pleasing results).

To create fences made out of thorny branches, or to allow nature to take its course? Regardless of which option you choose, one thing is clear: if you need to quickly create a windbreak for your horses while at the same time satisfying their need to chew, you know what you need to add to the paddock.

If you want your Benjes hedge to be full of butterflies (hummingbirds love many of these as well!), green your fence with plants that have no known toxic effect on horses and are popular with multiple species.

Some suggestions to get you started:

- Coral honeysuckle—not to be confused with invasive honeysuckles! (Spring Azure butterfly, Snowberry Clearwing moth host plant)
- Goldenrod (valuable host and nectar plant for multiple species, crucial fall nectar source for migrating Monarchs; do note that there is one species—rayless goldenrod—that *is* toxic to horses and must be avoided)
- Passion flower (Gulf Fritillary—only host plant, Zebra Longwing host; do note that this has a relaxing effect on horses, and is a banned substance in USEF competition)
- Gayfeather or blazing star (Painted Lady, Swallowtails, Monarch nectar)
- Joe Pye weed (Skippers, Azures, Tiger Swallowtail, Monarch nectar)
- Beardtongue (Chalcedon Checkerspot host plant)
- Aster (Checkerspot, Pearl Crescent host; Monarch nectar)

- Dill, fennel (Black and Missouri Woodland Swallowtail, host)
- Tickseed (nectar for multiple species)
- New Jersey tea (Spring and Summer Azure, Mottled Duskywing, Eastern Tailed-blue host)
- Sunflower (Monarch, nectar)
- Ironweed (American Lady, host)
- Beebalm (Swallowtails, nectar)
- Pussy willow (Viceroy and Mourning Cloak, host)
- Arrow-wood viburnum (Hummingbird Clearwing moth, Spring Azure Butterfly host)
- Spicebush (Spicebush Swallowtail, host)
- Cornflower (Painted Lady, host)

Since these plants aren't just beautiful, but also useful, our paddocks, pastures and gardens can be divided by green zones which encourage biodiversity. With a little initiative, this "deadwood" hedge can almost immediately become a "living" fence: a rich source of food and habitat, while also enriching the soil. The plants you place here, with the help of decomposing wood in the lower layers of the hedge, will thrive.

Try planting even blackberries or raspberries—these blooming vines won't just decorate the hedge, but also strengthen it. Sunflowers can also add beauty while providing food for insects and birds. With this method, each parcel of land can have truly beautiful and unique fencing.

Other undeniable advantages of a Benjes hedge: they're inexpensive to build, and do a great job of shielding the area from dust and noise. All the benefits of a planted hedge can be realized with this system, but you won't have to wait for your plants to grow!

Willows take root and grow easily, but need plenty of water.

Woven Fences

Originally used for erosion prevention or stabilization of river banks, a woven fence made out of willow (our native willows are shrubs or multi-stemmed small trees, not to be confused with European and Asian imports like weeping willow) or hazel rods can be used as an element in gardens, yards, or even your pastures. These fences are very popular in permaculture-type farming. They can be living—created by planting cuttings that are woven together as they grow—or panels made by weaving dry branches together, and either way, it makes great use of willow and hazel trees. Most varieties of these plants won't harm your horses; even if they escape over this barrier, whether it's dry or green, it won't cause a call to the vet.

Adding natural panels to the pasture or paddock can add a refreshing touch. It's easy enough to buy finished woven panels made out of willow or hazel branches, but if that's not enough for you, willow grows beautifully from cuttings if given consistent moisture.

Soak your cuttings at least overnight before planting in early spring. Don't wait too long between cutting and planting as the cuttings will gradually lose their ability to root.

Weaving a Green Willow Hedge

Pound acacia stakes about 20 inches deep into the soil (or straight willow rods—sometimes older thicker branches of the same type look particularly good after the rest of the fence turns green). Between these, weave the willow branches. Starting at ground

Between the hazel fencing (which separates part of a garden from the pasture) and the electric fence, the owners plan to add a hedge.

All the willows in this photo grew from willow cuttings planted around the pasture.

level, thread the branches through, alternating whether they pass in front of or behind the stakes, from each side of the stake. The spacing between the stakes shouldn't be much more than 2 feet—larger gaps won't provide enough support to withstand a strong wind.

Sowing Willow Branches

Dig a trench approximately 20 inches deep.

If your land tends to be drier, you can add a layer of branches 6–8 inches deep to serve as a kind of moisture retainer.

If your land gets enough rain or holds enough moisture, you can simply plant the willow twigs about 12 inches deep.

You can plant several ways depending on your taste: perpendicularly for classic growth, diagonally planted (at an angle of 45° against each other to create a stitch pattern), or you can even plant both ends of a cutting in the ground to create an arc where buds will form to create a thick and interesting fence.

Don't forget—you can braid willow cuttings together immediately after planting if they're long enough, or you can wait and weave together the following year's growth.

Also, keep a close eye on the soil—willow needs a lot of water after planting, as well as plenty of light (forget about trying to plant this one in a shady spot).

If you take good care of your living fence during its first year, it should start to grow beautifully. Sooner than you realize, you'll be able to enjoy the long green shoots from your hard work. In the following years, you can cut the growth however you like; willow can easily tolerate even a hard prune. In addition to willow, you can also use hazel or dogwood plants.

What Does the Law Say about Fencing?

Every state has its own laws about the fencing required to confine livestock, and they can vary a great deal. In fact, while some states require fencing to confine animals on owned or leased property, others still follow an "open range" policy where property owners who want to keep livestock *out* must fence their yard. I'm guessing it's safe to assume you'll want to keep your horses confined to your own land and will plan to erect some sort of fencing. With that in mind, I recommend familiarizing yourself with local laws prior to any major construction of new fence lines.

So, while building fences that don't offend the authorities or the neighbors, let's take advantage of a system of natural fences which will keep bees buzzing, birds chirping, butterflies spreading their wings and all kinds of small creatures scampering...

As our horse's caretakers and trainers, we have to love nature and have plenty of empathy—so we should find it easy to care about the other animals who are part of our local ecosystem and understand their need to find sanctuary around our homes and pastures. I know this is a new way of thinking for many, but I believe that on horse farms, it can easily catch on!

chapter XVI.

STONE AND STONE WALLS

Stone is a wonderful natural material that can be both a functional and beautiful element when used as part of our construction plans.

Marking property boundaries and dividing spaces intended for animals, reinforcing surfaces or laying out footpaths—these are just a few of the ways natural stone can be used. Although highly decorative, stone is also very functional: it traps heat during the day and releases it at night, for example.

Stone has been used for thousands of years: to build not just simple walls, but also complex defense systems and homes. In areas where more stone was available than wood, there's a long tradition in stonework and its use for all kinds of construction.

Stone fences and borders were mostly created around fields, as farmers would collect stones turned up while plowing. These were eventually used to demarcate the edges of fields and roads, and to secure slopes.

Why Stone?

Stone has some hidden and remarkable uses. Apart from aesthetics (its variety of colors, shapes and shades) and its easy workability with concrete, there are several benefits to using stone.

Stone heats up in the sun and as the temperature cools at night, it transfers the stored heat to its surroundings. A stone wall can help plants near it to overwinter that would otherwise struggle to survive in open spaces. Spiders, lizards, and amphibians love to warm themselves on a stone; that is, all the creatures a horse owner should love to have on his or her property. Why fight pests with chemicals when nature herself can do the job?

The more of the animal world we can invite to inhabit our farms, the better chance we have of keeping pests from getting out of hand because there are plenty of their natural predators around. I love a stone wall, since under these walls, water condenses for small things. These walls are sometimes the only thing in a large, flat pasture where earthworms, for example, can find needed humidity on a dry day.

Last but not least, stone—unlike wood—is unaffected by wet weather, hail, heat, or wind. If stone is exposed to never-ending rain, it won't rot or split. Instead, it will grow a lovely cover of moss and fern.

Categories and Construction of Stone Walls

A) Dry stone, or dry-stack, using earth in joints where we want to plant.
B) Wet wall, or wet-laid, where planting is done in front, behind, or on top of rather than directly in the wall—or places designated as future green space are built directly into the wall.

Dry-Stack Walls

As the name suggests, these are built without mortar or concrete. The wall relies on the weight of the stone itself for stability. Because it will constantly shift during the year, an unstable bed needs to be used to allow the wall to move. Instead of a solid concrete foundation, you'll need 8–16 inches of compacted gravel, or sand and gravel as a base. The good news for beginners: a dry-stone wall can be more forgiving of mistakes than a mortared wall.

For lower walls, the first layer of stone should be half-way buried in the ground. For any wall higher than four feet, the first layer should be almost to completely beneath ground level. The width of the base should be about ⅓ the height, and it should be built with a slope of 10–20 percent (the greater the slope, the greater the stability).

Stones used for building dry-stack walls can be anything from quarry stone (smaller, flatter pieces) to huge irregular boulders. Quarry stone can also be placed upright but for this to work, it should be framed with larger, more stable stones.

An enclosure with a beautiful stone wall in Spain.

If you keep pure-bred horses, you can certainly appreciate the beauty of shapes and the interplay of individual parts of a whole. If the result of all your hard work is truly beautiful, it doesn't matter if it's a living creature or a stone wall. Both speak volumes about your efforts.

Such a wall will also need a carefully placed crown which, among other things, protects the vertically placed stone from things like the eroding effects of snow and rain.

Walls made from quartz boulders are seen from time to time in Europe, especially in northern Germany. They're often seen dividing hedges from grass sod. In these situations, it's necessary to monitor the growth of plants: although the roots of bushes can strengthen a wall's foundation, if left unchecked they can grow into the wall and tear apart the base.

The stones should be carefully laid out using a template and a carpenter's square to maintain the desired slope. Individual stones should fit together so that they don't wobble at all or protrude too much after settling. It may be necessary to trim stones on the visible sides or ends.

Rule of thumb: 30–35 square feet of visible wall surface corresponds to approximately one ton of natural stone.

If you don't want to interfere too much with the natural shape of the stone, it's impossible to avoid using smaller stones and aggregate to fill in gaps as the stone is laid, in order to achieve the necessary stability (absolutely nothing should wobble!) and desired slope. The only other option is to cut or shape the stone.

Be careful to avoid having the seam between stones align with those in other layers. That is, make certain to break the vertical seam—lay stones to form a "T" shape seam as often as possible. If too many seams line up, the wall will be weak and could collapse. An expert recommended to me that during construction of higher walls, always check that the wall is level after three courses (layers), give or take. This is the easiest way to keep a level horizontal plane.

He also warned me against placing stones vertically—vertical gaps are more susceptible to the elements than the spaces between stones laid on their longer sides. While building, it's important to constantly check the appearance of the wall as well—it should look continuous and even (unless you're going for a different look!).

The upper part of the wall, or crown, needs to be made with the widest possible stones to avoid creating gaps which allow water to leak into the wall—freezing water inside a wall is a huge problem!

Mortar Walls

If you don't feel up to the task of building a dry wall to exacting specifications, so that the result is not only stable, but can hold up to the elements and animals, you'll want to consider using mortar and concrete.

In a nutshell: you'll need to lay a stable foundation of high-quality concrete which goes below the frost line. Next, follow the principles for building a dry-stack wall, but each stone will be connected to the rest using mortar. I'll give you the advice I was given: use a mortar mixture with a 1:3 ratio. Keep the look of the wall "clean" by stopping short of the last half-inch to inch of the visible edge of the stone when spreading the mortar. Incorporate the necessary number of expansion joints (approximately one every 15–20 feet) for the length of your wall.

Expansion joints are important! A gap must be created to allow for the moving and "stretching" of different parts of the wall. As temperatures fluctuate, volume and shape change in all directions.

Gabion Walls

Another kind of stone wall is a gabion wall. This building style originated more than a century ago in Italy. The name is derived from the old Italian word for a large cage: *gabbione*. In simple terms, these walls are created by filling a metal frame with aggregate. They were originally used as protection for riverbeds and as retaining walls on slopes. Later, they were also used as noise barriers.

Gradually, they have made their way into more residential areas. They are particularly useful on narrow plots, where height without width is needed.

The boundary of this piece of land ends with an ivy-covered gabion wall. The horses don't have constant access to this area, but they do move through it on their way to being tacked and ridden. The owners have noted that the horses don't eat the ivy; in time, the entire fence will be overgrown. It's up to the residents here to decide whether that's the effect they want to achieve or not.

A gabion wall, like a dry-stack wall, serves as shelter for small animals.

It's fast and easy to build gabion partitions. The welded mesh is connected using strong wire spirals usually in the shape of blocks (although other shapes are possible). They tend to be galvanized to protect against corrosion.

The great advantage of gabion walls is that there's no need to build drainage—water simply flows out. Even extreme rain and frost don't overload these walls.

Even though ivy makes an easy cover for a gabion wall, it's better not to plant it around your horses. Although they don't usually tend to eat it, it's poisonous. Because of the speed of its growth and its ability to grow on virtually any surface, it's very difficult to keep under control once it's established.

Stone and Horses

In the days when riding horses were kept primarily in stalls, and only allowed out for short times, they could often perform acrobatic stunts which were beyond what's considered safe once released from their stalls. Being kept in small paddocks, it was of course much easier for horses to injure themselves as well. Because of this, it was common practice to clear paddocks of absolutely everything. To be considered safe for a horse, it had to be a completely barren, level area.

So much has changed in just a few decades! Horses nowadays are quite regularly allowed to spend most of their time outside. That outdoor space has changed quite drastically, too. Horses now have access to larger fields where they can encounter changes in natural terrain. The luckier ones have access to slopes, ponds, bushes, and sometimes even forest. Today, areas for horses can be constructed with deliberately incorporated stones and gravel, artificial ponds, (carefully protected) edible shrubs, and useful trees. Their space can be supplemented with logs, branches, trees, stumps, and stones of all sizes. The stonier the riding areas, the more reason to get a horse's hooves acclimated to it. This can be done by creating varied footing in basic paddocks. In addition to clay and grass, there can be sections with fine sand and gravel, pebbles, and large flat stones, too.

If the unsightly tape were replaced with wood fencing, the natural corner in this basic paddock could be truly lovely. Notice the unused feed trough which has been turned into a water station for bees and other insects. On hot days, this hosts a lot of bug activity. I would suggest a few more pebbles added to the bottom to prevent bees and butterflies from drowning.

Planted Walls

Strategically-placed plants in a stone wall can be a real eye-catcher. All types of stone partitions can be planted—some directly in the gaps between the stones, others in the crown, others adjacent to it and using the wall for support.

Along massive retaining walls that are meant to hold sloping terrain, you can plant a climbing vine. If the goal is to make the wall appear lower, ornamental shrubs can be added at the base, or overhanging plants can be added in the crown. Both options "reduce" the size of a gigantic stone wall and create a more aesthetically pleasing impression.

- **A classic dry-stack wall** hosts drought tolerant plants—the gaps in these walls can't hold enough moisture for anything else to survive.
- **Walls built with concrete** can, along the lower part of the wall, be connected with mortar only on the interior—leaving space in the gaps for planting succulents and other heat-loving perennials.
- **Gabion walls** aren't just easy to install, but make it easy to be creative about the many ways they can be used not only as a partition, but also as a shelter for small animals and insects.

Important Tips

- If your new wall is in a pasture, it should be protected by electric fencing for at least the first year to allow newly planted plants on or near the wall a chance to establish themselves.
- Check and double-check the potential toxicity of anything you're planning to plant around your animals. Not just the plant, but its flowers, fruits, and seeds. Sometimes even a small amount can be risky.

You can build, pave, and decorate with stone—it can be used in so many ways around your property. Much to my husband's chagrin, I bring home stones I collect from our fields and pastures. His mood always improves, however, when he sees it being put to good use around the farm—as a hitching post, a place to sit, a wall around the feeder in the basic paddock...

The incorporation of stone walls into the landscape is well worth the effort when you consider the benefits.

Stone is often used in zoos to protect tree trunks from the residents.

A mound created from the excavation of a trough (one that isn't used by the horses—they go right around it; I built it to water a raised bed), combined with manure. It's a great solution for catching rainwater and runoff from the paddock. The nutrient-rich runoff is fantastic fertilizer for flower beds, or if you like, this washed-up mud can be used to add to your mound. Before it dries too much, it can also be used to build a vertical "clay" wall surrounding the hillock.

chapter XVII.

MOUNDS AND HILLS FOR HORSES

Raised beds don't have to be used only for gardening or planting flowers; they can be put to use as-is for horses to provide stimulation and entertainment.

If you were to watch horses in their natural environment, you might notice that part of their journey is often a trip, as I like to call it, to the observatory. That is, to the place with the best view of their surroundings. Horses have a natural need to seek out better views. We can help them satisfy this need.

Building a Hill

If you decide to do this without an excavator, take your time. Gradually build up denser soil. You can mix in some manure as well. As you pile the material, let the horses work with you. Sure, they may drag some away or "help" dig, but with a bit of patience it will become a big, heaping hill. You can use stones, logs, or soil-filled tires to outline the base—anything with enough weight to help maintain the shape is welcome.

This is an optional element which you can easily do without, but keep in mind that a bored herd that doesn't have access to grazing will appreciate the distraction. If you build the hill outside of summer (when gardening and mowing probably take up far too much of your time), your horses will probably be off the fields and in their basic paddocks, and will be more than happy to help you out!

It's important to be sure that the horses trample and compress the pile, but don't cause it to collapse. Therefore, after adding a layer of manure, be sure to add a good amount of clay or stony soil. Over the next few weeks or months, use a shovel to pile the earth back up over your base (this will be easier work if you used something stable and solid to maintain your foundation). Once the hill starts to hold its shape, you can remove the added material.

If your neighbors are looking to find takers for unwanted soil—grab it! Pile it up where it's accessible to the horses. You'll probably create a lot of excitement, especially for the younger horses. Don't worry about their safety—they'll trample it solid.

When the mounds become "softened" with manure, horses may decide it's a good place to poop or lie down. This may be a good time to fence the hill off and use it to plant something like pumpkins or zucchini. When you do this, it may be time to make another hill for the kids...

And presto—raised garden beds. So much more than just a pile of dirt!

Hillocks and mounds in the basic paddock can add variety and encourage play. They provide your horses with an elevated place to monitor their surroundings from.

Horses shouldn't find either pumpkin or zucchini plants palatable. With a small fence (either out of just branches, tape, or electrified tape—depending on the horses), they shouldn't feel inclined to climb into the bed. Of course, this all depends on how much space your horses have: if it's a small area with only a pumpkin patch, they'll definitely find a way to entertain themselves with it.

If you decide to go ahead and build your horses a "lookout" in their paddock or a pasture, choose a place with enough space. Horses love to be able to run up and down hills—so, given the chance, they'll bring themselves back around after running off in an arc. You may even catch your horses playing "king of the hill."

If you throw plenty of hay on the hills, that can help your horses' overall health and fitness, especially if they're lazy and would otherwise skip the climb. On the other hand, don't force a sick or elderly horse who may have difficulty moving to climb for their food—make sure forage is readily available elsewhere.

Most importantly, don't forget to enjoy your horses' joy!

Ornamental and edible pumpkins thrive on compost or manure.

Young horses enjoying play time on the developing mound. Note that the tires used here to define the boundary have been well stuffed (with clay, sand, or gravel) so that the horses can't get their hooves stuck in them.

This pumpkin patch is thriving in a stone-lined bed of horse manure and wood chips. This photo shows the same hill as the photo on page XXX—it's easy to see why I'm enthusiastic about gardening with manure!

chapter XVIII.

WATER AND THE BASIC PADDOCK

Since 1990, climate scientists have warned us that the planet is warming—and the northern hemisphere is warming faster than the southern. Predictions of gradually warming temperatures have unfortunately been proven true, and changes are taking place much faster than early models indicated. One of the main problems we're facing already because of this: unpredictable rainfall and periods of drought.

Climate change is alarming, and as horse keepers, deeply affects us. It doesn't matter if we're recreational riders out in the countryside, or competition riders who rarely leave a manicured arena: we all need hay for our horses. And we need a lot of it. Without enough rain, the price of this precious commodity will soar, and prices have already begun to climb.

As responsible landowners, we can be a part of the solution. It's imperative that we help restore our soil's ability to retain moisture: parched and compacted land will simply not absorb anything from a heavy rainfall, especially as rain patterns become less reliable. Instead of a landscape of bare and trampled dirt, we can plant trees, hedgerows, and blooming flower beds which are fertilized organically with our composted horse manure. We can construct small (or large!) retention ponds to catch runoff and rainwater, and recycle this water to irrigate our fields and water our plants.

Why Is the Soil Overheating?

A flat landscape with no rain and no trees, with straightened rivers that can't meander (which are necessary for the creation of wetter ecosystems and water retention), without enough healthy forests, with too much pavement, with harvested fields left without cover crops... all this contributes to the overheating of the soil. Hot soil not only loses its ability to retain moisture, but at temperatures above 100°F, there is a sharp decline in the health of soil microorganisms. On a day when the air temperature exceeds 90°F, it's easy for the surface temperature in a bare field to measure 130°F!

In the past, abundant rain was taken for granted and areas were dried for agriculture. Excess water was diverted in order to increase production. No one could foresee the problem this has created for us today: a landscape that water runs over as if it were waxed paper.

If you happen to live in an area where this isn't happening, take a moment to appreciate your good luck before you begin the work to help your land stay this way!

On a harvested field which is left bare, the soil temperature can be 50° to 70° degrees higher than on a similar surface planted with cover crops. Temperatures measured by thermal cameras were lowest on wet surfaces, followed by forested areas, then unmown meadows, then mowed meadows. The highest temperatures measured (after asphalt and concrete surfaces) were taken from soil with no bodies of water or vegetation.

There are countries where water is so scarce that what you may consider the common practice of hosing off a horse is an unimaginable luxury.

Soil Compaction

Soil compaction can occur thanks to one of two factors, or a combination:

a) **Inherent characteristics:** Determined by the properties of the soil (for example, soil that is simply naturally dense and heavy).

b) **Anthropogenic:** Caused by human activity, especially by using heavy machinery in unsuitable conditions (on wet soil).

What Causes Soil Compaction?

Repeated foot or vehicle traffic increases density and reduces porosity of soil by pushing air pockets out. The basic biological and physical properties of the soil are damaged.

Soil compaction goes hand-in-hand with water retention loss and leads to soil deterioration.

The result is soil with a low capacity for water retention. Over time, even water infiltration (absorption) can almost completely stop. On a flat field, water will pond on the surface for days after a rain. On slopes, water run-off carries away topsoil, further decreasing the ratio of arable land. Plant root systems can't break through the lower layers of compacted earth, affecting growth and productivity (consider this in terms of grazing areas). Microorganisms also suffer declines in compacted soil.

Air temperature and humidity are affected by transpiration (evaporation). The higher surface evaporation is, the better the chances are for a functional water cycle. For this water cycle to work for us and our land, steam vapor from plants and the soil must be captured and cooled by leafy plants.

The presence of vegetation is crucial for a productive water cycle.

This cooled vapor is then returned in the form of dew or rain. If there's only a small amount of plant life to intercept this vapor, it's carried off by the wind. This often results in condensation over the oceans, where it's no help to our small farms! A small-scale water cycle, on the other hand, keeps water where it's needed and helps reduce extreme differences between daily temperature highs and lows.

The report from the UN's panel on climate change mentions sustainable agriculture and sustainable management of soil resources many times. What does the panel mean by these terms, and how does this relate to us as horse keepers?

Sustainable Horse Keeping in Rural Settings

If we define the term "sustainable horse keeping" as a farm's ability to house, feed, exercise, and train animals without negative impacts on the surrounding environment, this gives us a vision for our management and building plan.

Every effort should be made to manage our farms so the natural cycle can do its work to maintain healthy, nutrient-rich soil.

In order to have a closed cycle, we as landowners must ensure that:

Soil nourishes plants » plants nourish farm animals » animal waste and decaying plants nourish the soil » the soil is healthy enough to nourish new plant life and continue the cycle.

This cycle should be permanently repeatable (hence the term "permaculture"), without exhausting the soil and its surrounding ecosystem.

It should enable more than just bare existence—it should create a healthy and thriving environment for our horses and surrounding nature.

You don't have to be a permaculture pro right from the start to begin incorporating the basic principles into the management of your farm. Whether in a stable or a backyard facility, it can be easy to keep in mind the most useful aspects: manure is a commodity, not unpleasant waste. Kitchen scraps and yard waste are a key part of the nutrient cycle. You can begin to compost, to plant native trees, shrubs, and flowers around your farm.

You begin to look at your land not just through the eyes of a rider or breeder, but through the eyes of a caretaker who sees the horses as one piece of the whole natural picture.

Permaculture is a comprehensive philosophy not just for land management, but for the correct use of natural resources and for environmental stewardship. It's a targeted effort to maintain a perpetual and sustainable natural cycle where natural resources aren't depleted by human intervention but instead are regenerated at the same rate they're used. Permaculture farming requires a change of mindset in relation to our entire surroundings.

Water for Horses: Waste or Necessity?

Climate scientists tell us that in the following decades, there will be an increase in the number, intensity, and duration of heat waves and drought. Recent years have already proven them right. Dry, dusty, buggy summer days with yellowing grass in the fields and languid herds of horses roasting in the sun... even those horses lucky enough to have some wetter areas with green grass are worn down by the heat. Where horses have access to shelter, they'll often spend long hours hiding from the sun rather than grazing. In an attempt to give their horses some relief, some owners rinse their horses off.

But in some areas, wells are empty and there are municipal water bans in place. Plenty of us are at work all day, and can't take a break for the sake of our animals, so giving them some midday refreshment is just a pipe dream, right?

By creating bodies of water for our horses, we can give them much-needed comfort on hot days.

Not necessarily—we can give our horses access to water in the summer, or even year-round. We can put rainwater to good use; rather than just letting it run off our roofs, we can store it for gradual release, or even divert it to an artificial pond. Landowners with man-made ponds report that

This stable has a waterer with a small "wading pool" for the horses to dip their hooves into.

they're very popular with most horses, and feel it improves their animals' overall well-being. Your horses will surely show you how happy they are with their new pond—many even prefer drinking from them rather than from their buckets. Building your horses a pond with a shallow, sandy approach, or even just a tiny wading pool, doesn't have to be an impossible task. Sometimes it can be a happy accident: a pond is formed by removing soil to fill a path, or by hitting ground water when digging corridors between pastures. If you aren't this lucky, you can build a pond by diverting water from run-off or using one of your wells. It may take your herd time to notice their new water feature, or they may take to it right away.

Either way, you'll probably notice your horses drinking more water (if the water is clean), without any noticeable increase in mosquitoes or other pest insects. Some folks choose to leave these ponds accessible to their horses in the winter: the horses have no trouble breaking the ice with their hooves and often prefer to drink the cold pond water despite the readily available water in heated buckets.

Some areas are already being impacted by decreased levels of groundwater, but it's still common to hear folks dismissing this: it's always been this way, wet years and dry years, no need to get excited...

But I've had the opportunity to observe the changes up close and in person, and I feel differently. I've seen the same place change drastically over almost 30 years: wells have dried up, as have streams.

Small bodies of water allow for more water retention in the landscape, allow for better growth of plant life, and facilitate the development of necessary conditions for maintaining biodiversity.

I ride through a section of forest which used to have a pool of water surrounded by lush vegetation even on the hottest days—it has changed before my eyes into a dry path that even the toughest plants have retreated from. Where my horse and I used to struggle through the mud, the path is just as dry and dusty as any other.

This body of water pictured here is very popular with most of the resident horses. Most of the initially hesitant members of the herd were shown by their companions the joys of a good bath. The horses also enjoy drinking from the pond—many prefer it to their fountains of tap water.

Here, the pond is available to the horses year-round. Even in the coldest weather, they have access to it. The horses don't go out on the ice, but some will break the ice at the edges in order to have a drink.

A partial view of a beautiful outdoor stable in Holland which provides the horses with access to a water feature.

Practical Advice for Building a Pond

- Except when the ground is frozen, you can begin digging your pond any time of year. The more the horses are able to "supervise" construction, the more willing they'll be to accept the new element in their paddock.
- Choose a site according to your pond's water supply: as close as possible to buildings if you'll be depending on rainwater from rooftops, and if you're using spring water, make your decision based on the abundance of the water supply. Whatever your decision, be sure it's an area your horses are already comfortable going to. If you construct a pond in an area your horses don't like to be, your pond will still be useful, but not to your horses.
- If you have a choice, don't locate your pond by deciduous or pine trees. These trees will greatly increase the maintenance requirements. If your pond isn't surrounded by bushes and trees, your horses will more readily accept it.
- Don't feed hay near bodies of water. You'll needlessly add to your cleaning work on windy days when piles of hay end up in the water.
- Although it's possible to fill your pond with water from the municipal water supply, you should carefully consider before resorting to this option. Horses prefer untreated water to chlorinated water.
- The bottom and sides of the pond can be lined with a thick layer of clay, a liner, or concrete. With concrete, there's a risk of cracking in the cold, so it must be very high quality and the right methods must be used for building. No matter what you choose as a base, it should be installed so that it won't injure or scare our horses. Which brings us to the next point:
- Gravel can be used as a bottom surface—it's fine for horses' hooves—but it should never be used together with a liner, due to the possibility of sharp stone edges tearing the liner. Safe options for backfilling with a liner are smooth, flat pebbles (never round, as these can roll under a horse's hoof), load-bearing anti-slip rubber tiles, interlocking rubber pavers, rubber impact plates, artificial turf, or commercial-grade carpets. Deep clay can swallow a lot of backfill materials, but most horses won't want to go into a bare clay pond (the gooey mud "swallows" horses' legs). If sand or fine gravel is available, it can be used to pack the clay sublayer, but keep in mind when digging that adding this will decrease the depth of the pond and dig accordingly. Never have a bare liner without backfill—if it tears or folds over, horses can injure themselves.
- Horses poop when they're nervous, so never force a nervous or frightened horse to enter the water. Clean up any manure that enters the water as soon as possible—although our horses don't have the same standards we do for clean bathing water, excess droppings will turn your beautiful pond into smelly sludge.
- If you're using plants to filter the water (a self-cleaning pond), make sure the horses don't have access to this (separate) part of the pond, and be sure you have enough vegetation to do the job.

Concrete Wading Pools

A wading pool made from concrete must be built with horse safety at the forefront: no sharp edges, a roughened surface so horses won't slip, or lined with a non-slip surface like rubber pavers, commercial carpet, or artificial turf. Unlike a deeper pond, wading pools must be drained in the winter to avoid injuries. Shallow water will freeze solid, and the horses won't be able to break it with their hooves—it will be far too slippery even during barely-freezing temperatures. If horses will still be able to access the drained pool, it's a good idea to cover it with something like sawdust or sand. Often, this helps it serve double duty in winter—the horses may choose to use it as their bathroom. If you have a larger wading area, filling it with fine sand can provide horses with entertainment during the winter months. It won't freeze solid, so your horses will have a place to dig and lie down. If your wading pool has a raised stone edge, it's best to block your horses' access to it when it's drained—horses can hurt their legs on the edges, especially while rolling.

Ditches or Moats

A water feature can add variety to a basic paddock. Where there's a path or trail that horses must travel through from time to time, part can be dug somewhat deeper to catch water. (Be sure it's done in such a way that horses aren't afraid to enter.) There's anecdotal evidence that this can help maintain the health of the hoof horn, and horses often seem happy to visit these areas...

Be sure your concrete wading pools are safe!

A moat right on the trail. If the horses want to use this path (they have other options—they don't have to), they need to go through water.

Swales

These are ditches of any shape or size which are built on sloping land to contain runoff from stormwater. The water isn't held here permanently, but soaks into the surroundings or feeds into a body of water. Over time, water saturation in the land below the swale (it will be drier uphill) will increase, giving surrounding vegetation a good water source.

If you want to combine a swale with planting, bushes and trees should be planted near, but not in, the ditch: it's occasionally necessary to remove sediment deposits from the swale, and bushes and trees would make this difficult to impossible. If your swale is in a basic paddock instead of a grassy area, you'll need to remove sediment deposits more often. If the sediment is sand, it should be returned to where the water took it from. Other deposits will be nutrient-rich and can be added to flower beds or compost.

A swale can also be a convenient source of water for nearby plants, which can be a gift in areas that go long stretches between rains. If the horses leave this area alone, the only real maintenance is the occasional removal of sediment deposits (which can be great fertilizer or be used as a "modeling clay" to reinforce raised beds).

This swale which I built at the lowest point of my basic paddock provides water for both soaking and watering.

If, however, your horses enjoy digging at the banks of this ditch, the banks can be reinforced, for example, with large flat stones.

The swale should be located in a less popular part of the basic paddock so that it isn't a dangerous addition. Horses should just pass through this area, rather than choosing to stay here. Since the basic paddock doesn't have vegetation to hold back runoff during heavy rains, you'll need to periodically return sediment to your paddock, especially if the ground is sloping. Much of the sand in your paddock can be kept in place with barriers (like tree trunks), with the remainder being caught quite nicely by the swale. (The photo to the right shows a divider made of tree trunks—since the horses need to step up to enter this defined feeding area, they don't track as much sand.)

If you decide to use a shower or sprayer, be sure to choose a water-saving device.

When the area is mostly dry and the swale is only somewhat moist or muddy, you may notice it attracting a wide array of birds: swallows will scoop up mud for their nests, sparrows and titmice will come to take dirt baths, among many others—pure joy!

Don't forget how easy it is to create a fountain or bath for birds outside of your horse areas as well—around your garden or home—just add water in a suitable place and regularly clean and top it off, giving you a lovely element to enjoy outside your window.

Sprayers

Another option for making a hot day more pleasant for our horses is to install a shower or sprayer. It can be installed directly into an existing pond (if the water is sediment-free), using a circulation pump to spray water. Alternatively, it can be connected to a rain barrel, cistern, or well (which is located outside of the horse's enclosure, so the water is kept clean). The main safety issue to never forget with water and horses: electricity.

The only difficulty the future will hold, it seems, are the water sources. Keeping this in mind, build with nature's possibilities in mind—use rainfall efficiently and avoid anything which needlessly wastes water.

Bodies of Water and Mosquitoes

Plenty of people see water near horses and have the same reaction: "Never! Water breeds mosquitoes!" Even so, these same people often have plenty of mosquitoes in their own paddocks.

What's Happening in Actual Practice?

This may surprise you, but most people who add water features for their horses notice no increase in insect populations. This is probably because in order for a water system to function properly, it needs to be cared for well: they plant plenty of vegetation to clean their pond, which in turn invites frogs to eat the bugs, and encourages dragonflies—who aren't just wonderful because of their elegant beauty, but are also great bug eaters, and their predatory larvae will feed on mosquito larvae, effectively eliminating any "bug problem."

Open rainwater containers certainly can be a breeding ground for mosquitoes, though. Make sure containers are enclosed and don't have a still, open surface.

Mosquito-repelling herbs and flowers to add around your pond:

- basil
- lemon thyme
- lavender
- marigold
- rosemary
- sage

The repellent effect of these plants is due to the high content of essential oils (which deter pests while attracting pollinators). Essential oils are released from the plants through evaporation and are unmissable to insects.

Even in this bucolic setting, the available water feature is a highly frequented place to drink, bathe, and just pass through. As you can see, there is plenty of space for planting a variety of plants.

What Does the Law Say about Water Features?

In general, be sure that you aren't diverting water from a neighbor's property or creating an issue with water draining onto adjoining property. As laws can vary from state to state, be sure to check with your local and state government regarding construction of any large artificial water feature.

Typically, no special permission is required for simple rainwater collection devices. These devices are simple to install and easily removable. In some areas, in fact, rain barrels are not only allowed, but encouraged (as they keep storm runoff out of the sewer system). Check with your town to see if they're available to you at a discount.

Every drop of water held a moment longer before it drains away counts!

Many horse owners teach their horses to enter a newly constructed water feature in a halter. For safety reasons, don't forget to remove it after the lesson.

Bodies of water are a regular feature in extreme-trail areas.

A Message from a Vet

If you have bodies of water available to your horse during the winter, be sure to prevent access to them during any event which may frighten your horse (for example, celebrations with fireworks). Animals frightened by loud noises may run onto the ice, even though they would normally avoid it. A horse who slips on ice can suffer catastrophic fractures. As winter approaches, consider not just the overall nature of your herd, but also specific high-strung individuals.

chapter XIX.

SCRATCHING POSTS AND SCRATCH PADS

One of the essential items a basic paddock should provide our horses is a place to scratch their itches! A horse won't hesitate to use a tree or bush, but not everyone has those in their enclosures. We have to give them other options, since they can't really scratch their backs or tummies on a wall or the edge of a shelter.

You can buy brushes and scratchers made for horses or cattle. Some horse owners are worried about their horses getting too much hair caught in the brushes, but horses with access to lots of trees and bushes leave far more hair tangled in those than on scratchers in their runs...

Always ensure scratching structures are safely installed!

If your horse is spending too much time scratching the base of his tail or the crest of his neck, be sure he's clean, has no underlying skin irritations, and is dewormed—even internal parasites can cause itching. Also be sure your horse's sheath or udder is clean, as this can also lead to tail rubbing.

If you're trying to protect a long mane and worry about using a brush with long and stiff bristles, you can also install a DIY scratcher, for example a door mat or foot massage mat. You'll need a solid wooden fence or wall, or a stake set well into the ground so the horses can't knock it over. Once you have a location, it's extremely easy to install a mat with a couple screws. Whatever hardware you choose for installation, be sure it's smooth: it shouldn't have any sharp edges that your horses can hurt themselves on while scratching.

You can also make a homemade scratcher by installing curry combs and brushes on the edges of shelters. I've even seen entire scratching stations made out of wooden pallets, which are covered on all sides with different sorts of scratchers. Be careful—these constructions can be so popular that you will probably have to build additional support and an excellent anchoring system for them.

Cattle scratchers can be excellent options for horses. Before installation, take a close look at the metal parts and, if necessary, sand any sharp edges into a safer, rounded shape.

A brush "tree." This home-made device (made out of crossbars and firmly planted in the ground) can make up for a lack of trees on your property or save the trees you have from excess wear and tear.

Brushes on the canopy of a shelter.

Never use nails to install a scratcher! When wood dries, nails tend to loosen and horses can cut themselves on a pulled nail.

Perhaps even more than with mats on stakes, plan for all possible problems when installing.

Frequently Asked Question:

Our horses use a wall-mounted brush to scratch themselves all over, but can't scratch their bellies—what can we do?

The horse's tummy is often forgotten. To help a horse get a good belly scratch, sink a tree stump into the basic paddock at a good height, and screw a massage stick or brush onto the top.

Ordinary brushes screwed on top of posts on homemade scratching stakes are easy to install. On these, horses can easily scratch their shoulders, the bases of their necks, and just behind their jaws. If you're designing a homemade scratcher, consider the small, itchy areas (like the spaces under their jaws, or their cheeks); a rubber handle, installed at the right height, can make an itchy horse very happy...

Don't forget that sometimes all a horse needs for a good tummy rub is a large, well-placed branch. This also gives him something to chew on.

Some folks are champions of invention. In Germany, I once saw a homemade system where flexible brushes had been installed into tires at different heights using concrete. The tires were then buried so that only the brushes were above ground. They were very popular for under horses' jaws, for bellies, and for "undercarriages."

This industrial universal brush, slid onto a homemade holder, has wonderful long bristles to relieve a horse's itching.

chapter XX.

TOYS FOR HORSES

Games with objects like this should always be done under supervision.

You may be thinking that the best toys for a horse are other horses and a large pasture with plenty of grass and bushes...

...and you'd be absolutely right. But, as with most things in life, not everything goes according to plan. Even a happy herd on a large pasture can have that one horse who won't leave the mares or foals in peace, or is always biting at the geldings, or chews on every fence post. What should you do, when the group can't calm him down and your horses are getting visibly tired of his shenanigans? Or what about an injured horse who has to spend weeks laid up in a small paddock? How do we distract these horses for a little while or give them some entertainment?

Manufacturers have created any number of toys intended for these scenarios. While some are intended for healthy horses living outdoors, others are for stalled or sick horses who need to be on limited movement. You can seek some of these out, or get inspiration from the homemade gadgets on these pages.

The most popular toy, often, is a ball, especially with young horses. If you have a fixed wooden or vinyl fence, you don't have to worry about a big ball rolling out of your enclosure, but with electric fencing, take care not to let the tape or wire break due to excessive play. This could also lead to chasing a horse who's chasing a ball—through broken fencing, and down a road...

A small percentage of horse owners feel very strongly that horses shouldn't be left with balls in paddocks or in the pasture, as there is a possibility that a horse could hurt himself by jumping on an unstable object like this. It's up to you to evaluate the risks of leaving your horses to play with a ball, or whether to choose a different game for them.

Edible toys are very popular, too, including: salt licks or other goodies suspended from cords, carrots stuck in hay nets where horses need to work to dig them out, apples thrown in a water tub, or, last but not least, twigs and branches for chewing.

This last one can often provide the most entertainment.

Take Care with Conifers!

Many pines, due to the high content of essential oils and resins (which can irritate the horse's gastric lining in larger quantities), should only be given to horses in limited amounts. In smaller quantities, however, these same things can be beneficial. Be careful not to introduce Ponderosa pine to any pregnant mares, as the needles contain an abortifacient.

This "Likit" is a salt-lick combined with a game. Be careful, however, not to allow a bored horse to consume excess salt, as this can lead to diarrhea.

If you decide to use toys and games only with your supervision and more for training than for fun (for example, getting horses used to different colors or noises, the movement of objects over their heads or around their bodies, and so on), balloons can be a great option.

Of course, horses left to their own devices will find their own "toys," such as drinking containers, feed buckets, and wooden structures. What seems to be most interesting to many horses quite often is large farm equipment, sometimes even when it's being operated...

Of course, it bears repeating: **the best toy for a horse is a happy herd!**

Of course sometimes the most popular toys...

...are the ones you don't voluntarily give them.

This treat can be long-lasting.

Extreme trail obstacle or a toy in the paddock?

chapter XXI.

FOREST GARDENS

Having horses right next to your house, in your backyard, doesn't mean you have to sacrifice aesthetics.

Nowadays, the idea of keeping two or three horses at home is gaining in popularity. It's not easy to create a lovely garden in an area used for these large, heavy animals, but with a little effort, it can be done.

One option to modify flat bare areas is to divide space and create little "habitat islands" with raised beds, hills and embankments. These can be combined beautifully with another useful and lovely landscaping option: forest gardens.

Lots can be done with even a relatively small space. Most of us, even if we have horses on our properties, want to have inviting, modern yards. If the yard next to someone's house looks and smells like a dump—full of broken equipment, trash, and manure—this gives all of us who keep our horses at home a bad reputation. I don't want my kids to be ashamed to invite friends over because of a stinky farm!

This is why everything we build should be done with the goal of beautifying the space and adding to the surrounding landscape. Any change made to the yard should help to maintain or enhance natural beauty, not destroy it or steal from it.

A New Way to Look at Land

Most of us need to think outside the "traditional" box of land management for horses, since we don't have unlimited space for them to trample. In order to successfully incorporate horses into a smaller space while also keeping it green and inviting, we can look for inspiration in a few methods, including the Paddock Paradise (or PP) system, and the principles of forest gardens.

We can create a functional and aesthetically pleasing environment for our horses by creating separate spaces which are connected by paths; each path is bordered by "forest islands" and fenced off. These paths, or "Paddock Paradise" trails, lead the horses from the areas adjacent to the home towards more open spaces such as pasture. Although the areas outside the fencing can be planted however you please because it's fenced off, I still recommend using non-toxic plants.

Keep the big picture in mind as you build: choose a uniform type of building material for things like hay and equipment storage, the manure pit, and so on. A sheet metal shed for tools would stick out like a sore thumb near a wood house and a wood-fenced paddock, but if it's sided with wood or covered by bamboo, it can blend in with nearby buildings. Likewise, if the house is surrounded by rose bushes and a rock garden, it makes sense to plant the spaces in front of the adjoining stable in a similar way. Set farther back or behind buildings, you can create a transition to native plants and woodland using carefully selected trees, flowers, and shrubs.

If you don't have a lot of energy to devote to this, choosing a few undemanding plants can help make life easier.

Solid paths are the basis for a forest garden for our horses.

For those of us that love to garden, however, this can be an opportunity to create something really lovely that can have long-term rewards.

Be very objective with yourself about how much of a green thumb you have. Nothing truly limits how much you can do, but if you think you may struggle, be very sure to choose easy-to-grow and easy-to-maintain plants that are suitable to your specific habitat.

Horses kept near a home make a lot of demands on their surroundings. Unless you have unlimited space, you'll definitely spend some time and effort combating summer dust, spring mud, and year-round smell.

One horse shouldn't have to live alone (an ideal group consists of at least three horses—although two is sometimes enough), but be careful not to antagonize your neighbors with more animals than your particular piece of land can handle. Once you have everything well-planned-out and have neighbors who are open to having horses next door, get to work!

Reinforcing Footing

Smaller high traffic areas by the home should be laid with sand and gravel, pavers or mats, or even artificial turf.

Keep in mind that in order to keep horses at home in a smaller space, reinforcing paths that are available to them year-round is a requirement, not an option. The areas by the home need to be both useful and look good.

If you decide to use gravel, be sure to combine it with other materials, so your horses aren't forced to stand or walk only on stone. They should always be able to leave gravel areas, not just to walk on varied surfaces, but also to lie down and roll. It's not easy to keep an area used by horses looking good, especially during wet and muddy seasons. Look around at what other horse owners in your area are doing to see what options may work for you.

Expect to use plenty of backfill. If you've already planned to set up a basic paddock and paths with any surface used for keeping horses outdoors, you'll probably have several loads of sand and gravel delivered to use as a foundation for that final surface layer. I can tell you from experience that even with sand paths, it's not just helpful but necessary to prepare a foundation. The soil underneath should be removed and replaced with compacted gravel (some add a liner underneath this) before adding the surface layer of sand.

It's no small job to build spaces for horses right in the backyard; plan well to avoid living with long-lasting mistakes.

Let's be tough on ourselves as we build our spaces. It can be tempting to start, for example, with a flimsy temporary shelter instead of a more permanent (and safe) wooden one to "give it a shot" before building something better. Before you know it, the temporary shelter you had planned to replace is still there years later, looking a bit rough around the edges.

A smaller area near the owner's house—the center of the field has been left for use by the family or for occasional grazing, and the perimeter has been reinforced with gravel and made available to the horses year-round. This path leads to the riding arena and to pastures adjacent to the family garden. You can see in this photo that it's mud season: the path isn't lined and stone was poured directly on the soil, making it necessary to add a new layer of stone from time to time (which has just been done). As time goes on, less stone should be needed.

Let's always keep in mind that the elements we incorporate should be safe and useful, but they should also look good. Be uncompromising about appearance: this is the only way to build near our homes so that the space can be a private little paradise for us and our horses.

It can be hard to be critical enough to get rid of things that aren't working any more for the space, but this is how you keep your farm from slipping into disrepair. It can be easy to convince yourself that the stuff that's somehow still hanging on is enough, but it isn't!

The paths in this photo have been reinforced with large rubber puzzle piece pavers and bordered with wooden ties. They can be easily removed at any time. Even though the cheery colors fade over time, they remain functional.

Try not to cut corners as you build. Maintaining high standards really pays off in the end!

Don't be afraid to have high expectations for the beauty around you. After all, we share this space with our horses. Not everything that suffices for them will also satisfy us.

If you have wide open spaces for your horses, you can stick to the traditional idea of open spaces which are always grazed or mown, but this chapter looks at a different concept: stables that are incorporated into forest space, or forest spaces that are built near our homes and connected to them by plantings.

The Fundamentals of Forest Gardens

Some people like to surround their land with taller trees, leaving all the interior space open, and others prefer to include trees in their yard, so they live with the feeling of being protected by the treetops. Although it all depends on your personal preferences, I recommend not only enjoying trees for their protective and ornamental value, but also for their use as part of the menu.

At this German stable, solid paths encourage movement from basic paddocks to covered areas for rest. Solid surfaces outside the narrow trails are generally spacious. Outside these areas, the stable also has traditional pastures.

Skilled craftspeople will look for species suitable for basketry (willow, hazel), for example, while others will think primarily about winter heating and look for trees that can easily be turned into firewood.

Frequently Asked Question:

What do I need for a forest garden? How large a plot of land do I need to have in order to consider doing this?

A forest garden is a system of mixed cultivation: trees, shrubs, and crops all growing together to maximize space. Together in the same space (which doesn't necessarily have to be large), animals can be kept. On a quarter of an acre, an edible forest garden can be functional and lovely. The essence of this concept is the interconnectedness of the individual components.

Owners with small yards need to connect different spaces! How else can they provide horses with a lot from a little, and still be able to grow something for the family? Can you see where the concepts of edible forest gardens, the paddock paradise system, ornamental areas, and keeping horses at home intersect?

Basic Principles

Get started by reading up—it's easy to find plenty of tips out there for how to connect a functional garden with ornamental gardens or forest. Unfortunately, it's much harder to find information about connecting this with care for livestock, and even less about keeping horses in this setting. However, many choose to do everything they can so their horses can be a part of a living system. When these horse owners turn to me for advice, I send them looking for professional literature. Horse ownership is very time-consuming and requires knowledge about the animals as well as several related fields (agriculture, horticulture, nutrition, veterinary medicine, saddle fitting, shoeing...). Did I scare you? Good! Because underestimating what is needed to keep horses at home only causes the animals to suffer. Common sense isn't enough: it can't replace hard-earned knowledge.

But enough threats—let's get into forestry! I won't address the basic principles of forest gardens from the perspective of crop and yield management (although I recommend becoming familiar with this interesting topic), but rather the best ways to maximize use of land so that gardens and areas around the home can be managed in order to provide an ideal setting for your horses.

Forest gardens are not a typical garden with a "useful" part and an ornamental one: they are wilder and more natural. They can appear almost unkempt to a casual observer (since we're used to looking at manicured and mowed front yards and symmetrical tree lines), but it isn't so. These "messy" gardens take years of painstaking labor, but when they're done, they can become almost maintenance-free and look very tasteful. The initial stages aren't easy and it can be tempting to throw in the towel, but if you stick with it, you'll be rewarded with so many benefits.

A lot of advance planning goes into incorporating horses into the design. My advice: leave your horses boarded elsewhere until you've laid all your foundations, built out most areas, and put in at least initial plantings. The moment you bring your horses home, your energy will be devoted toward their care and not things like flowers and vegetables.

My next piece of advice: add every element to your forest garden, including fruits and vegetables, spices, herbs, edible leaves, materials for your fencing, and firewood.

Your little paradise should be a place for all living creatures: birds, bees, butterflies, groundhogs, frogs, and more. Bordering the trails, plant habitats for wildlife. Feel free to add winter bird feeders and puddling stones for butterflies in the summer.

The hardest thing about maintaining backyard horses in a small space is balancing necessary animal welfare with the aesthetics of their surroundings.

You're probably wondering how all this can be included while maintaining a functional space. I mentioned the concept of the "Paddock Paradise" at the beginning of this chapter: a system of paths and marked trails. Vegetation will be behind fencing where it can grow and thrive, and the horses will be on dry, reinforced paths between these gardens.

I'm not recommending planting every possible square inch and forcing your horses to live on narrow paths! The essence of the PP system is that these trails lead somewhere—to a forest, an open sandy area, a hill with a view, or a pasture.

Hollyhock in a paddock: this plant is considered medicinal and is safe around horses but is listed as invasive in many regions in the US. Consider native alternatives like blazing star or bee balm.

Protect tree roots from the destructive effect of horses' hooves! You can lay perforated mats over them, which will add cushioning but still allow water to easily soak through.

Paddock Paradise in More Detail

Everything depends on how much space you intend to leave for your garden, and how much you'll leave at your horses' disposal. With this as your starting point, you can plan in more detail.

Start by choosing a good spot for a shelter. Leave a square in front of it at least the size of a round pen (15 to 20 meters in diameter); anything larger is a bonus. Plan to reinforce the footing in these spaces. If you make the base sand, this can be a suitable place for ring work as well as for horses to enjoy at liberty. On the opposite side, create a second area. If, instead of sand, you have mats in front of your shelter, this opposite area should have sand. If you have space for a third area—forest—you have the foundation for a complete paddock paradise.

Connect these spaces to each other with paths; splurge on larger pastures, if you have the room. The minimum recommended width is 10–12 feet. Personally, I feel like a little (or a lot) wider is better, but definitely don't go below this minimum width! Ideally, your corridors shouldn't lead to a dead end, but should create a travel circuit from facilities (shelter) to rest (straw and mats for lying down) and play (forested, areas with licks and games)—with hay along the way so the horses have to move to find food.

Think about protecting trees—not just the part above ground, but also the network of roots. Otherwise, your forest won't serve you or your horses for long...

A small plot near the house that serves horses and people. Outside the barn, there is a shelter and a basic paddock. A paved path surrounds the entire property. Horses are only allowed on the inner pastures for allotted times, but they have hay available year-round. The owners plan to add garden plants around the space, creating an area for the horses to move about freely and for their people to enjoy gardening. This plan to add a living "edible fence" and more useful trees will help create privacy from the neighboring development. The trees growing in the basic paddock will eventually offer an option for outdoor shade (so the horses won't have to go in their shelter if they prefer not to).

The ideal PP system has at least one path leading to more open spaces, like a pasture. Under no circumstances should we confine our horses only to the paths—keep this in mind during planning and construction.

Once the space is planned, the trails are built, and the horses have access to different "zones," all that's left is to plant around the horses and the home.

Don't believe that in the PP system the animals live only on paths—far from it!

This final piece is a bit challenging if we decide to add a forest garden. The concept of this kind of garden is to create a fully functional ecosystem.

Trails around the pasture and garden. To the right and left, since it's a large plot of land, vegetable beds and edible forest can be established without the owners having to give up pasture space. In the lower right corner, part of the steps leading to the pasture are visible. This entire path, which isn't yet reinforced, is currently being maintained with a layer of gravel separated by the soil by cloth, all covered with a layer of sand.

Everything here has its role: even ornamental plants should be planted with an eye to the benefit they bring to the ecosystem as a whole.

Forest garden construction includes careful consideration of rainwater management, not only by collecting water in containers and reservoirs, but also by modifying terrain to slow runoff. This, together with mulching and composting of manure, create a system where horses and plant life can support each other, rather than compete with each other.

Characteristics of Forest Gardens

- A sustainable long-term community of plants that provide nutrition for the landowners
- An ecosystem that should require only minimal maintenance after the first few years and should be able to weather seasonal changes
- Plant species ensuring biodiversity

How Do We Make This a Reality?

Envision a lush garden where the vast majority of plants are edible. The plants are combined so they don't compete with each other, but grow symbiotically. They are grown in at least three layers: herbs, shrubs, and trees (but more is better!).

Try to imagine that under the apple and pear trees, an edible ground cover of blackberries and raspberries, chives, and sages grow instead of grass. Bushes and shrubs bear fruit under the edges of birch trees. The path leads you under a grove of nut trees, where shade-loving strawberries flourish.

With a little bit of research and work, even a small piece of land can be used to create a garden for yourself, your horses, and an array of insects and birds (who will thank you by taking care of all sorts of pests).

Rosehips along the trails—the fence is high enough to protect the bushes, but low enough to allow the horses to harvest the fruit all autumn.

For your horses, you'll create an enriching environment that provides much more than just food and shelter. For yourself, you'll create a little piece of paradise. And a lifetime of physical fitness.

Forest gardens ideally consist of:

- Layers of tall trees (pine, pecan, oak, birch…)
- Layers of lower trees (fruit trees, American hazelnut…)
- Layers of shrubs and taller herbs (blueberries, echinacea…)
- Layers of perennials, vines and herbs (parsley, native honeysuckle, bee balm…)
- Ground cover (strawberries)
- Underground layers (carrot, potato, echinacea roots, ginseng)

A park-style forest garden.

Here, a space for ornamental planting has been left between the trail and the house. This photo was taken during the initial stages. Later on, we would see a colorful cascade of flowers and lush grass.

When establishing "bio corridors" around horse areas or trails, it's ideal to combine woody plants that offer protection and shelter as well as nutrition. To be successful, layers of plants are necessary, while also taking into account the appropriate placement of sun-loving and shade-tolerant understory plants.

Choose vegetation according to the growing conditions of your particular region. The lower layer of grasses and herbs under the shrubs should tolerate trampling and shade—you'll be working around the shrubs and trees not only to harvest but also to care for your new plants. Don't forget that a carefully constructed windbreak around the trails can block winter snowdrifts.

Although it's not an easy or common style, this use of space works and is worth the hard work; it's up to you to decide whether you're ready to do it. Imagine relaxing in your garden, enjoying the sight of verdant gardens and grazing horses and the sound of birds singing in the trees, everything functioning as a part of the whole... wouldn't you like to live like this?

Forest gardens are a great way to combine a love of horses and gardening.

chapter XXII.

GIVING HORSES SOMETHING TO CHEW ON

Take a look at any wooden shelter or fence at a stable, and you'll notice one thing pretty quickly: gnawed boards, beams, and walls. It seems like a mystery that some structures are untouched while others have been mercilessly chewed.

Experienced horse keepers have systems in place for preventing damage to fences and run-ins, since they know what a horse who can't be out on grass is capable of.

Especially in winter, when our horses are in their basic paddocks or in fields with no grass or other plants and only hay to eat, their life can be a bit boring (even if it's lovely hay). After they finish eating their hay, they'll probably take to exploring their surroundings for something else to chew on. Chances are they won't have straw bedding or another distraction, so eventually they'll notice their feeders (if they're made of wood), a fencepost, or their shelter. Posts or corners can be wrapped with sheet metal or fine mesh (which can serve double duty as a scratcher). Another option is to cover these areas with (non-toxic) branches, which can be eaten instead of the structure itself. Anything that takes the horse's attention away from expensive wooden shelters and fencing can be a helpful addition.

In Europe in the 1700s, cattle and horses were often allowed to graze in forests. This meant that farm animals ate not just grass, but shrubs and other forest growth. Dairy cattle that were provided with this combined grazing produced much higher quality milk. Although this was wonderful for the animals, it was devastating to the forest undergrowth: saplings were stripped, and cleared areas couldn't regrow.

What Does This Mean for Our Horses?

For the horses we love, it means we should work to satisfy their innate need to graze and forage. At the same time, we need to consider the damage to the land that these large animals can cause. If we're going to be responsible caretakers of both the animals and the land they're using, the health of the trees on our property should be as important to us as the health of our horses.

If we don't want to see the depressing sight of dying trees in our paddocks and pastures, we must either:

- Provide a small number of horses with a large piece of land which has plenty of trees and shrubs
  or
- Come up with clever solutions to preserve vegetation on a small parcel of land

This tree serves double duty as a feeder. The hay dropped from the bale protects the tree's roots from the horse's hooves, and the hay distracts the horses from the tree bark. Last but not least, the large number of woody plants in the pastures keeps the horses from chewing the tree.

Paradoxically, more horses can be accommodated with the latter option than the former: we'd never expect 5 acres of land to stay lush and green if we put eight horses out to pasture year-round without any restrictions...

Protecting Wooden Buildings and Trees

- Discourage animals from chewing by coating surfaces with a non-toxic deterrent spray.
- Wrap beams and trunks with mesh, textiles, metal strips or similar materials (but make sure that whatever you choose is harmless to horses!).
- Provide horses with wood specifically intended for chewing and gnawing.

Don't be fooled: just because something is common practice doesn't mean it's the best practice. A common way to keep horses from chewing on the wrong things is to provide them with all-day access to hay, but this doesn't take into consideration a horse's need for variety. Our horses miss having things like branches, herbs, and old grass when it isn't available to them (during winter).

If you have a large area with some forested spots that you can use for your horses, you may read this and think I've lost my mind. You're probably wondering why anyone would make life so hard for themselves, pulling things out of the woods to put in the paddocks when the horses can just be let loose among the trees. Although you may own a beautiful, large pasture with just the right variety and quantity of trees and shrubs and varied terrain, many people don't have this option, or are simply protecting their land.

How Much Is Too Much?

Wild horses have always foraged on buds, leaves, twigs, and bark as a small but important part of their diet. Some types of trees are safe for even our domesticated horses to nibble on. The question that remains: how much should be provided or allowed?

So, although some may think adding branches to the basic paddock is totally ridiculous, it's a good option for a lot of folks who want to preserve the health of the trees on their land.

With many things that horses eat, the amount eaten can be the difference between a health benefit or a health risk. Horses that are kept outside year-round on a large area with access to a variety of forage (not just grass, but also other plants, shrubs, and trees) will usually have the instincts necessary to respect the thin line between enough and too much. Some trees and herbs, for example, are medicinal if small amounts are eaten, but can cause poisoning in larger doses. We can assume that wild horses instinctively honor the rule of "testing" unknown plants by only consuming small amounts. Moreover, most poisonous plants, when fresh, usually have an unpleasant or bitter taste which horses have no trouble avoiding. The problem is often not with obvious

Mixed forests that are excessively grazed by livestock can't self-renew.

These structures and trees have been protected by wrapping them with wooden branches that horses can easily gnaw on. Using this system, damaged branches can easily be swapped out for new ones.

toxins, however, but with "somewhat" problematic plants: those that can be beneficial in small doses, but dangerous in larger amounts.

This is where horses need to have excellent foraging instincts, passed along from experienced members of the herd to the youngest (often from mothers to foals). Horses who live outdoors year-round on varied terrain can learn selection through this cooperation.

Unfortunately, this isn't a learned behavior among our stabled horses, who have always been turned out in areas with only a meager offering of plants or with no vegetation at all. Horses who live mostly on hay don't have the opportunity to develop the instincts necessary to evaluate and choose appropriate forage while grazing, and it isn't clear to what extent this disappearing trait can be relearned, if at all, in a single generation.

Today's modern horses often lack the ability to self-regulate when it comes to safe levels of food intake. When this is combined with boredom, it can be lethal. Health complications often arise from

These horses have been provided with plenty of branches to enjoy.

Tree bark can be an excellent source of nutrients and minerals, and can also provide valuable phytochemicals that can aid with digestion and with consistency of manure in winter. In the summer, it can help to balance blood sugar levels and to maintain weight. Young buds, leaves, bark and small wood branches, together with hay and pasture, can provide all the mineral needs for horses if the necessary variety of species is available.

excessive consumption of what should be a harmless food (for example, fallen fruit).

Many of our horses are kept on pastures sown with only one or two grasses: no herbs, shrubs, and no "old grass" during the winter. Often, they literally have nothing to choose from—they're given a monotonous diet of single-species hay and single-species pasture. Many people don't even know what their hay supplier is bringing them. These practices aren't only impoverishing our horses' taste buds, they're also harming their health. This type of forage, sown only for its yield and not its composition, can't provide all essential nutrients, even when growing conditions are ideal.

This soft-stemmed hay also fails to do anything for a horse's teeth. Nature provides tooth care to wild horses in the form of hard fodder like branches, bark, and tree trunks.

We can help our domestic horses compensate for some of these deficiencies by introducing branches (from non-toxic trees) into their enclosures.

Any plant's nutritional content will depend on where and what time of year it was grown. It's also important to keep in mind that nutritional value is not the same as the nutrition available or usable. This makes it impossible to say with any certainty that there's a precise feeding ration for woody leafy matter. Content may vary not just across the seasons, but also year to year, area to area, and plant to plant.

In addition, each horse will have differing abilities to extract nutrition from their food. When we start giving branches to horses who haven't been given this type of forage before, we can't necessarily assume they'll have a good sense of the amount they can eat or the pace they should be eating at. Start with very small amounts of non-problematic species.

If you decide to use old tires as root or trunk protection for a young tree, keep in mind that they can become far too small for the tree after only a couple years. This problem can be solved by cutting the tire. Another issue with this method: it won't work for anyone who wants to honor the natural beauty of the land.

We can draw a lot of inspiration from functional enclosures in zoos. In this photo, you can see how branches have been placed so they're available for gnawing, but also so that they protect the tree trunks.

Only after a period of habituation, and after we've made sure the horse's initial enthusiasm has faded, can other species be added.

The goal of every horse keeper should be to maintain a balance between the nutritional value of woody plants and their function as part of the landscape. We've already discussed the value of windbreaks, plant belts, and hedges. I'll just point out that to have horses at home in the idyllic environment we all envision, it's not just knowledge about horses that we need, but also horticulture, gardening, and landscaping. So, study, consult with experts, and—most importantly—plant!

New plantings must be not just cared for, but also protected—from horses, other animals, children, and sometimes our own overzealousness. Maybe you think caring for horses is more than enough responsibility and you don't need to also be running around taking care of plants. Try to look at it this way: you're not just supporting the environment, you can also enjoy fewer trips to the supermarket by growing fruits and vegetables. If this makes you feel like rolling up your sleeves and getting to work, keep reading!

If we want to be able to cull our trees and bushes so we can give some branches to our horses, these plants need an undisturbed living space that is protected from our horses. You can learn to love and protect your gardens just like you love and protect your horses: they are also alive.

Giving Horses Branches to Chew On: Some Rules

- Only branches which are safe and suitable are put in the paddocks.
- Always check leaves for mold and fungal diseases.
- Always start with a small amount and a limited variety. Greater quantities and more variety can be added after horses prove they will only eat reasonable amounts and won't stand over the branches all day.
- Branches are never used as a substitute for hay rations; they're only added in sufficient amounts to relieve boredom and supply crude fiber or nutrients.

St. John's wort—an example of a poisonous medicinal herb that we don't want to see on horse farms. It can be dried and hung in bundles under the roof of sheds, however, to repel insects.

- Old branches are taken out when new ones are brought in. There should always be enough space in the paddock for horses to safely move.
- If you're working on the construction of a Benjes fence system, old, chewed wood can be left in place and covered by new layers of branches.
- Never use branches from any tree that has been recently sprayed or otherwise treated with pesticides.

- *Never* use horses as trash receptacles: don't get rid of rotten or moldy fruit, or diseased branches or leaves, by throwing them into the paddock for your horses.

I'll discuss in some detail the apple trees you'll have planted in your garden—as you prune them, consider sharing the branches with your herd. Most stone fruit trees (cherry, plum, peach, apricot) are potentially toxic to horses, as are pear trees; other varieties (gooseberry, currant) are a possibility in Europe, but are banned from the US.

Apple Trees (*Malus domestica*)

Apple trees pose no risk to horses. Their wood, bark, buds, flowers, leaves, and fruits all provide horses with valuable bioactive compounds, which, if you observe the guidelines listed previously when making branches available to your horses, can be nothing but beneficial. All kinds of apple trees are safe as a whole—you can give your horses a branch complete with leaves and fruit. If your horses aren't accustomed to daily access to apples (they're not turned out in a pasture that has apple trees that produce fruit), the fruit itself should be given in limited amounts (this is true for any fruit—not just apples).

Throwing garden scraps and extras into the paddock can be a major health risk for horses. You run the risk of veterinary bills for medical complications like colic that can cost thousands of dollars, or, worse, losing a horse.

Fruits

Studies have shown that apples can have a beneficial effect on a horse's gut microbiome. Healthy horses can benefit from even a small amount of fruit added to the menu. If your horse's diet needs to be managed due to health issues, however, consult your veterinarian before adding anything to his diet.

Apples are rich in pectins (which, among other things, inhibit the growth of harmful microorganisms in the intestines by binding toxic substances), as well as citric and malic acid—all of which have countless health benefits.

Leaves and Bark

The leaves and bark of apple trees contain phenolic compounds—mostly flavonoids. Phloridzin, the primary flavonoid in apple bark and leaves (and in lesser quantities in the fruit as well), is a powerhouse: it protects against damage from UV radiation, strengthens cell walls, and reduces blood sugar levels.

Don't forget to leave some fruit on the trees—animals and birds will appreciate it.

A planter with herbs available on the horse's path: the protective grid ensures that the horses can't overgraze or uproot the plants once they're well established.

It also acts as an antioxidant with benefits for cardiac health. In addition to all of this, it has antimicrobial properties, fighting off *Staphylococcus aureus*, *Listeria monocytogenes*, and *Salmonella typhimurium*. This flavonoid is present in the bark, fruit and leaves (it has also been found in the bark—but not the fruit—of pear trees); the fruit has the least amount. The leaves, on the other hand, can have a pronounced effect on blood sugar levels.

The common plum can be a problematic tree for horses. All parts of the tree contain high levels of amygdalin (although somewhat less in the fruit). These levels can vary from year to year and even from season to season. This is why a horse (and possibly only one out of a group) can go from having no problems with this tree to suddenly being poisoned. Sensitivity to amygdalin can vary from animal to animal, and even change during the lifespan of the horse. We can never forget that each horse also has different appetites, leading one horse to consume far more than another might.

Only a healthy and uninjured tree can provide life-giving energy to its surroundings. A sick tree, on the other hand, will drain energy from its environment. Keep this in mind when preparing to allow horses into new areas where trees are thriving. The trees should be protected from the horses because of the benefits they can offer them (and all living things).

Buds

Compared to leaves and bark, apple buds have a significantly higher concentration of phenolic compounds (antioxidants). Unfortunately, in order to have a successful harvest, we can only feed a few buds to our horses and only for the short time of year when they're present. Leaves are a much better option for expanding our horses' diets, as they're available for many more months.

Blossoms

Apple blossoms tone and strengthen an exhausted organism, so they're wonderful for healing and convalescence. They strengthen the liver and kidneys. (Even humans can enjoy these benefits: steep plucked flowers in boiling water for 15 minutes, and then sip slowly throughout the day.) Always check that the flowers you choose are healthy, not dry or brown.

chapter XXIII.

INDIVIDUAL MEALS AND THE BASIC PADDOCK

Keeping horses outdoors has its pitfalls. One of them: giving grain or other concentrated feed in the paddocks or pastures. Feeding two horses is easy, but trying to feed ten or twenty—better put them in stalls, right? Not necessarily! In this chapter I'll discuss the gadgets that have proven themselves and the tactics owners have successfully used to feed the horses where they live—outside.

Those in the know emphasize the need for peace at meal time: unlike hay, horses get very excited about food served in buckets, since these are rationed meals they only get at certain times. Horses that are excited or otherwise stressed while eating run the risk of esophageal blockage. Unfortunately, owners can unintentionally "help" with this...

The term "trough feed" is nowadays increasingly referred to as "bucket feed" in outdoor stabling systems.

Beet Pulp: Soak, Soak, Soak

Beet pulp is a popular option, but it wasn't always this way. Made from the leftover material after sugar is extracted from sugar beets, this used to mainly be given to dairy cows. Nowadays, however, you can find it in most barns. It's an excellent source of digestible fiber—it's sometimes given as a hay supplement when yields are low, and can be a great option for horses that have trouble chewing hay. Nutritionists often like to include beet pulp in their equine clients' diets due to its excellent digestibility.

Where does beet pulp come from? Sugar beets are washed and cut into thin slices resembling potato chips. The sugar is then extracted through a procedure involving heating, steaming, and soaking. The pieces of "pickled" beet are then pressed to remove excess water, and the pressed fiber is dried immediately to prevent spoiling. After further heating and drying, the resulting low sugar material can be chopped up into what you and I would recognize as beet pulp.

This chopped-up material contains—among other things—soluble carbohydrates and insoluble cellulose, lignin, and pectins (soluble amino acids are leached out during sugar extraction), leaving a low-fat and low-mineral content, although it does have a relatively high share of copper and manganese. The cuttings are fairly palatable, and due to their high concentration of pectins, they are beneficial to a horse's digestion. Pectins are characterized by their ability to bind water and form a gel that settles like a film on the surface of the intestines. This creates a favorable environment for beneficial intestinal microflora.

Pectin—a polysaccharide classified as a soluble fiber—is found to reduce cholesterol in the blood and can also trap heavy metals. It's easily fermentable (which is to say digestible) in the small intestine.

Beet pulp contains little starch and nitrogen, but plenty of digestible fiber. Some manufacturers recommend soaking it in water for a minimum of 9 hours at a ratio of 1 part beet pulp to 4 parts water; others recommend 10 to 12 hours at a ratio of 1:5.

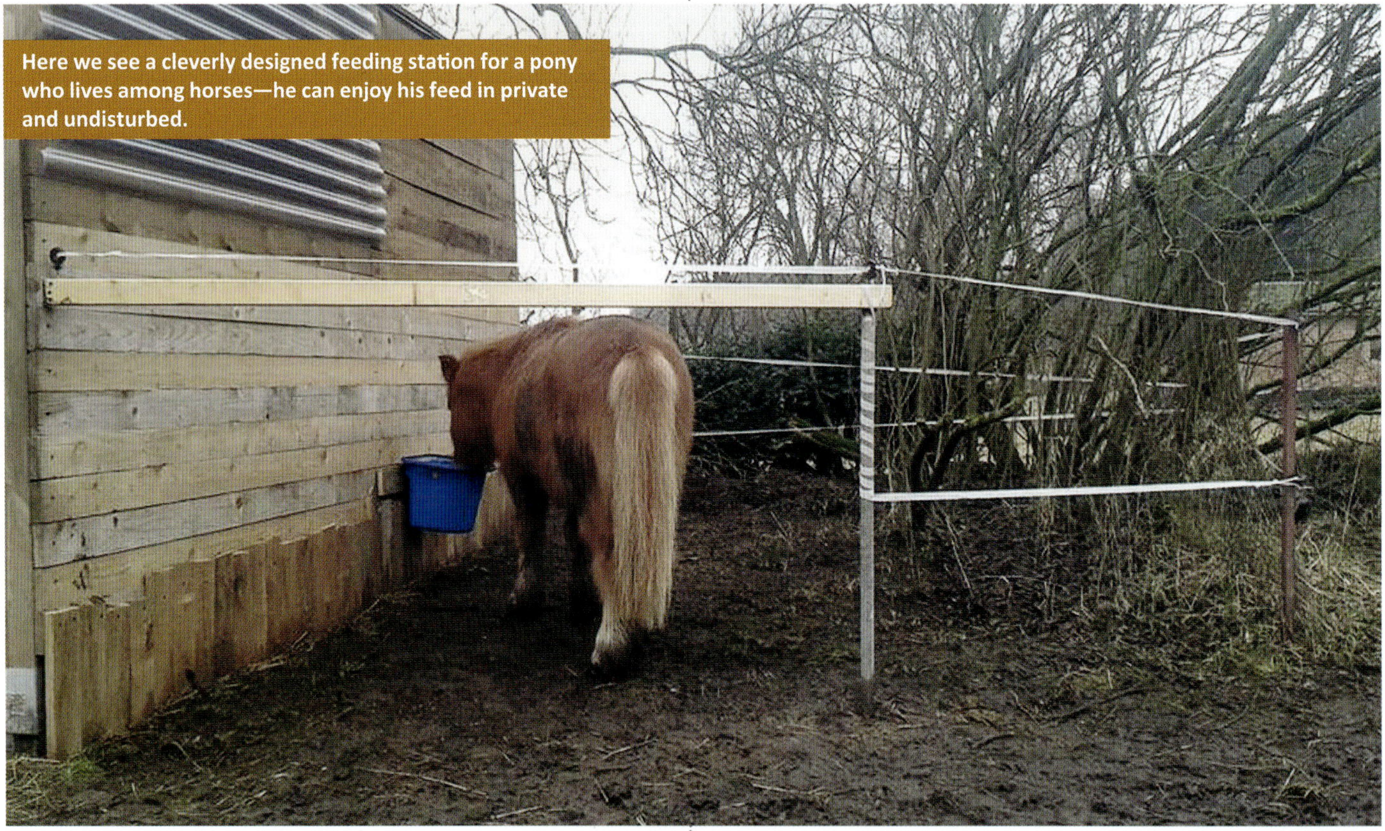

Here we see a cleverly designed feeding station for a pony who lives among horses—he can enjoy his feed in private and undisturbed.

Pectin fiber, with its many benefits, has one huge downfall: it's highly expandable and can cause a serious blockage in the horse's esophagus. The most common secondary complication is aspiration: liquid or particles are inhaled into the lungs, where they can cause pneumonia.

To quote Barbara Bezděková, Ph.D, DVM: *"Intraluminal obstructions of the esophagus are, among all esophageal obstructions, the most common and are caused by impaction of the esophageal lumen with food (hay, a bite of feed, and so on). This type of obstruction often occurs with a voracious intake of feed, or in the event of dental abnormalities or sedation which can cause incomplete chewing of feed. Another contributing factor to esophageal obstruction is the administration of dehydrated feed (for example, beet pulp) without prior soaking, which, after mixing with saliva, leads to expansion of the feed. A spasm of the esophagus plays a significant role in causation. Affected horses should be denied feed, because some horses (young horses especially) will continue to eat even through an obstruction until they have obstructed all the way to the pharyngeal region."*

It's clear from research and observation that apart from changing or incorrectly preparing feed, another variable also plays a role: gluttony.

Why Do Horses Gobble Their Food?

This happens most often due to stress, or more precisely because of worry that another horse, or even a person standing near them, will deprive them of food. Lifting a leg while eating and pawing while waiting for grain (or while eating) are symptoms of increased stress levels. To make matters more complicated, it may not be stress that causes a horse to gulp his food, but excitement. This excitement can cause a rush of endorphins that, with some horses, wears off very gradually—so this rush will last for the entire time the horse is eating.

Some horses tend to tie their emotions to the amount of grain they get (less grain = more stress). A horse that used to be a happy eater, suddenly put on a diet, becomes anxious at meal time... There's a simple solution: add chopped soft hay to their ration of grain. I recommend a DIY approach to this; store-bought mixtures can sometimes irritate horses that may be prone to ulcers.

If you're feeling fancy, you can adjust the feed volume in a bucket of grain by adding, for example, diced carrots or maybe freshly-picked dandelion leaves (cut into small pieces).

To introduce the appropriate solution, the triggering factor for stress during feeding needs to be identified.

One thing is certain: if we can manage to create an environment where a nervous eater can be calm at feeding time, the risk of this horse choking on his food will be greatly reduced. Although some horses are anxious about protecting their grain, others may become stressed if taken away from the herd, even if it's for a yummy meal.

Ensuring the well-being of any particular horse when feeding an entire herd is not an easy task. A wild horse, even if he has to search and scavenge, has food around him all the time. This horse doesn't wait for a bucket of feed to come to him twice a day. He doesn't have to worry that he'll miss out on something special if he's a bit slower than the rest of his herd. Having a rationed amount of food that arrives at specific times can create a real risk of choke with a nervous horse.

Let's look at a couple methods that can work well in this situation.

Option 1: Spacing Buckets

One of the most popular methods in use today (assuming most horses eat similar amounts of the same feed) is to carry buckets out and space them at distances that work for each horse. The horses are left loose and are fed in the order of their established hierarchy: the boss first; after him, the second in charge; then the third; and so on. This way, the horses go gradually to their individual buckets, and,

Not every herd can calmly eat from the same trough without fighting.

ideally, eat calmly. The first horse to finish his grain will probably push the nearest neighbor away from his bucket, if he knows he can get away with it.

However, if most of the horses get the same amount of grain, most will finish around the same time. If the buckets aren't taken away immediately, the final carousel usually takes place peacefully with horses passing by empty buckets and licking the last crumbs from a neighbor's pail before going along their way...

This method is very easy with two horses (even if they get different grain). It's also not a bad option for a smaller, close-knit herd with a well-established herd dynamic, if they all get similar amounts of the same grain. With a larger herd, there must be at least two people to dole out the meals to avoid having the first horse finish while others are just starting to eat; otherwise, one horse will have two meals while another has none...

If some horses need extra feed or they get medicine or particular supplements, the horse's owner or caretaker can sort this out when that horse is being worked with individually. Or, if just one or two horses in a group have special needs, those individuals can be separated within the enclosure or pulled from the pasture for meals.

Disadvantages

It may take some time for your horses to adjust to this routine, and during this adjustment period, there may be a few "explorers" who walk away from their buckets to see if anyone else has anything more interesting... This is a good time to stand watch between the horses. Some horses learn the rules pretty quickly (don't take a step toward those other buckets!), while others may only become more nervous and make everyone

else nervous, too. Another problem can be tipped buckets and trampled feed.

Tips

To Slow Consumption

There are plenty of tricks to employ to get horses to stay at their buckets longer—by adding chopped forage to the bucket of the horse who would otherwise finish first, or by tossing treats around the buckets (like carrots) so horses don't immediately harass their neighbors.

If a horse likes to toss his grain out of his bucket, you can add a few large (smooth!) stones to make it difficult for the horse to toss mouthfuls of grain. However, some horses may get so fed up with this that they simply upend their entire meal—keep an eye on the behavior of any horse you try this trick with, since the goal is less stress at meal time, not more!

Built-In Feeding Stations

Some horses are just playful—what can be done with the horse that's more interested in playing with his bucket than eating what's inside it? I recommend a fixed trough or bucket. In my paddocks, I have heavy earthenware troughs sunk into the ground at all the raised edges of the paddock, ten in total. They've been in use for years without any problems. They're deep and wide enough that they could be buried sufficiently to prevent even the most industrious digger from moving them. Of course, it helps to discourage digging by putting the troughs near the electric fencing, so the horses have no real interest in sticking their hooves in the area (but not so close that the horses can't toss up their heads while eating). Even young horses will very quickly stop playing with something that doesn't move, roll, or break easily.

"Indestructible" Buckets

If you prefer using buckets to investing in troughs, look for materials that are guaranteed to last longer, like buckets made of rubber. There hasn't been a bucket invented yet that can survive a horse's attention if it isn't picked up after meals, though. If you want to leave buckets in the pasture indefinitely, there are no guarantees. Finally, always pay attention to health safety: be sure whatever you choose is made out of food-safe material.

Option 2: Feed Bags

In Germany, it's common to see horses being fed with a feed bag or bucket attached to their heads. There's anecdotal evidence that most horses adjust easily to this method—but there are always exceptions. Some folks prefer this method for feeding, since each horse can have a meal tailored to his specific needs and they can't go after each other's meals. With this option, it's important to acclimate each horse individually before using it with the herd as a whole. There can be a moment of panic when a horse raises his head and the bucket doesn't stay on the ground. For anyone interested in introducing this method to her horses, please take the necessary time to teach them to be comfortable with feed bags.

Option 3: Tying

Other horse keepers deal with meal time by tying their horses. This method has two benefits: first, each horse stays at his own bucket and we can be sure each horse eats exactly and only what they were given, and second, the horses learn to associate being tied and standing quietly with something pleasant. I recommend untying horses more or less right after they finish eating. If this time is used to try to teach horses to stand quietly after they've finished their meals, they will become impatient and anxious to be unclipped. You can still use this as an opportunity to take a moment after they finish eating to check them over, pick their hooves, and give them a tummy scratch before they're released. If they don't have to wait long, they'll learn to associate being tied with both feed and standing quietly.

How to Tie the Horse

Always start by having a halter, either leather or rope. You'll be tying the horse to a ring that has been installed on a fence or wall. This ring then has a length of twine attached to a carabiner or panic snap. A thin strap of leather can also be used. Although it won't give immediately, a panicking horse will be able to free himself.

Don't Rely on a Panic Snap

You can't rely on a panic snap to release when a horse panics. Most of them are designed for easy, one-handed release by a person: they won't give if a horse is pulling on them. Those that are designed to open if a horse pulls often only work if they are yanked. A horse that slowly increases pressure (some will even lie down) won't be able to free himself without help.

If you tie a knot rather than using a clip, make sure it's easy to untie even if the horse is leaning his full weight on it (see illustrations below).

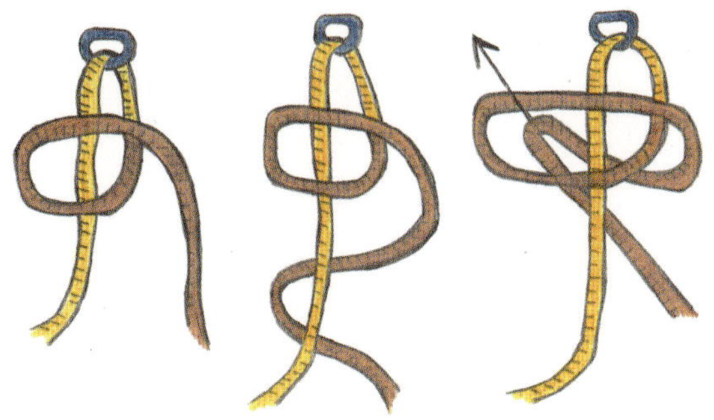

Where to Tie the Horse

- **To a shelter.** Be sure it's sturdy and well anchored. A flimsy structure won't handle a panicked, tied horse. I've seen a brick wall torn down by a tethered horse—and it happened in the blink of an eye. No one had a chance to untie the horse.
- **To a solid fence in your pasture.** Check the wood regularly to be sure the wood hasn't rotted.
- **To a tree in the enclosure.** But be sure the tree won't be harmed!
- **To a beam.** Add an intermediate piece of twine or leather to tie through as a precaution.

Never tie to:

- Benches, tables, trash cans or any other insufficiently anchored object. A horse who falls or runs backwards can flip even a very heavy object...

If you can't think of a starting point (or you have a place in mind, but it isn't well-anchored), try to build a special shed that you can really rely on. In the photo below on the left, you can see a very nicely constructed solution of tying to a beam over the horses' heads. This can be a safer option than tying at shoulder height, where a horse can get a leg over the tie. The space is divided into good old-fashioned straight stalls. For a horse that requires more space when eating, leave one space free and a wall on the other side. This isn't usually necessary; most horses get used to feeding quietly and calmly if there are partitions by their heads.

Spilled Buckets—A Common Problem

Of course, no one likes it when buckets are spilled, tipped, or tossed into the stratosphere. Hanging buckets can be knocked over if a horse puts his head underneath... supplements are expensive, medicines even more so. Even grain itself isn't free. No one wants the meals they prepare for their horses to just get dumped.

How should you deal with this problem with your outdoor horses? It's possible to find large feed containers (for grain and also for hay, with a slow-feed grid) with wide bases and low centers of gravity that are harder to tip. Or you could try a DIY approach like putting a bucket into a tire with a similar inside diameter.

The bucket-in-a-tire method can work for a lot of horses, but a more clever horse can still find a way to pull out the bucket and toss it.

People who have had to deal with endlessly playful and creative horses have come up with a lot of ideas for securing buckets out in the pasture. For example, a hollowed-out tree stump or section of tree trunk (partially buried in the ground) can be a hard-to-tip pail holding option. Some take it even farther by putting a tire (which is higher than the wood) around the tree trunk, making it nearly impossible to tip or remove the bucket. Another benefit: horses are less likely to paw at their bucket than if it were just on the ground.

What if you simply want to have a spot where you can hang buckets? It's a good idea to build at least one solid wall within the electric-fenced enclosure where buckets can be hung.

If you don't want to collect buckets after every meal and would rather have a more permanent setup, you can consider something like this: a long trough feeder made out of two halves of a plastic barrel. This can be used for grain or hay. The feeder in the photo has gaps between the boards, but I recommend building something without any spaces, since a horse can get his hoof caught.

Feeding pens are an absolutely acceptable option as well—you can build permanent pens or just use electric tape. If you'd like to try your hand at making something more sophisticated than a simple electric tape enclosure, the photos below may inspire some creativity.

If you use feed pens, you may find that you won't need to close them in from behind because feed time will be calm and orderly.

These colts are using metal feed barriers that can be adjusted to the necessary width for the horses using them.

chapter XXIV.

FEEDING STRAW—YES OR NO?

A question that sometimes comes up is whether or not horses can eat feed straw. For many people, it may not be clear what the term "feed straw" actually means. This term is often incorrectly understood to be used as a quality assessment: dry, undamaged by mold or moisture, minimal dust and no inappropriate admixtures. Feed straw isn't necessarily suitable for feeding to our horses.

Grain Straw

The term "straw" refers to the leaves and stems that are left after grain is harvested from any cereal crop (grains, legumes, oilseeds). Since we're focusing on suitable feed for horses, we'll only look at straw created from grain harvests.

"Feed straw" isn't a designation for a quality of straw (the term is not an attempt to claim it can be used as feed, not just bedding); instead, it indicates grain variety. It must be straw that comes from either grain sown in the spring, or grain that was sown the previous fall and has overwintered. Feed straw is never harvested from winter grain; due to the longer growth stages, straw from winter grain is far less digestible than spring straw. Keep this in mind if you don't want your horses to colic. When you see the term "feed straw" in literature, now you'll know what it actually means.

Barley and oat are the varieties of feed straw that can be suitable for horses.

Oat straw is believed to be the most palatable, but in practice, horses tend to prefer barley to oat. This can make oat straw a good choice, since they shouldn't eat it too quickly. Oat straw is considered the most suitable variety due to its slightly lower fiber content and its higher mineral content, as compared to other straws.

Barley straw is usually preferred by horses; some may even choose it over a poor-quality hay. There is literature that mentions some varieties of barley having a sharp hull that can be dangerous, but the vast majority of these fall off in the fields or remain on the grain. This concern can easily be avoided by looking for a hull-less variety. I personally have never had any problems arise from feeding barley straw.

Wheat straw is highly absorbent and less palatable, making it most suitable for bedding. If this straw is used as a supplement during a hay shortage, take extra care. This feed straw is the one most likely to cause colic or blockage.

Characteristics of Straw as Feed

Straw is classified as a bulky carbohydrate feed. It has a very high dry matter content (over 85 percent) and a high fiber content (over 30 percent).

Straw is a fantastic material for bedding horses outdoors. Anyone who is able to constantly bed with straw (and clean up soiled bedding) will have happy horses and healthy pastures. Straw is a fantastic addition to the manure pile, aerating it and balancing the ratio between carbon and nitrogen. If properly stored, your straw-filled manure pile will become wonderful compost and fertilizer.

Due to the high lignin content, straw is far less digestible than hay.

In a feed ration, straw can be beneficial as a source of coarse fiber, for mixing with hay to slow intake for our fast eaters and our curvy critters, and to help with boredom (horses have something to occupy themselves with when they're finished with their hay).

Straw should be added gradually and carefully to the diet of any horse. Water intake should be closely monitored, especially during the habituation process. Crushed straw shouldn't be used as It can increase the risk of constipation. If it is used, it should only be given to horses that are accustomed to straw, and it should be fed as an addition to long-stemmed hay. If at all possible, avoid feeding it entirely. It's wonderful for bedding, but feed only uncrushed straw.

After horses are acclimated to it, the amount of straw given can be gradually increased up to the stage where they have access to an entire bale and are free to forage as they please. In winter enclosures, be sure that lower-status horses don't find themselves in a situation where they don't have access to anything but straw. It's best to mix both hay and straw in all the feeders and hay nets, so no horse is left without access to hay.

Some horse keepers report laminitic episodes in their animals after feeding straw, but after caring for horses for more than thirty years, I've never encountered this. The difference may be that I know exactly what straw I'm feeding and how it was harvested. I also never serve it in large amounts or without accompanying it with hay. Moreover, I provide it year-round without any fluctuations.

I hope it's completely unnecessary to state that moldy or musty straw should never be used as forage or bedding, especially not in an enclosed area. Somewhat inferior quality straw can be used to dry out an outside spot where horses lie down, but it's important to note that I'm talking about slightly lower-quality straw, not a dusty, moldy mess. Furthermore, horses should have access to plenty of hay and good straw so they're not tempted to eat straw used for bedding.

The same rules apply to the feeding of straw as hay: after baling, wait approximately 6 weeks to feed. The cleaner the straw (no green remnants of weeds or grass), the sooner it can be used after baling. Clean, dry, "dead"-in-the-field straw with no green plant remnants can be used as bedding immediately

There is a big difference in the amount of straw a horse will accept if they have it only in their fields versus mixed with hay in their feeders. Feeders gradually acquire a high value in the minds of our horses: eventually they'll be more willing to eat something out of a trough or hay net that they might not eat off of the ground. Because of this, the only thing that belongs in a horse's feeder is high-quality roughage.

chapter XXV.

KEEPING HORSES WARM AND COMFORTABLE IN WINTER

Horses are naturally equipped to handle the colder months; summer heat tends to cause them more trouble than winter temperatures. As soon as the days begin to shorten in the fall, hormones responsible for coat growth kick into action by beginning to produce longer and thicker hair as well as a thicker undercoat. In addition to daylight, temperature plays a role: thermoreceptors will perceive a cold autumn and trigger the growth of a thicker coat. If a horse is kept indoors or blanketed, horses adapt by producing less winter fur.

As soon as the days begin to lengthen in December, horses will stop producing warmer coats, even if January brings extreme cold. This isn't something to worry about, as horses who have spent the seasons in the same place will have a well-established coat—even a sudden hard frost won't be an issue. (This is why it's best to transition a horse from an indoor situation to your outdoor stabling in spring or summer.) However, it's important to have well-fitting blankets on hand in case a horse develops a problem regulating his temperature due to illness. I'm not a fan of preventative blanketing for outdoor horses, but this doesn't excuse irresponsibility. Each horse should have at least one insulated blanket and a waterproof sheet, even if they both gather dust for years. At the end of the day, it's better to have an unnecessary blanket in your closet than to have a shivering, wet horse out in a blizzard. Calling a vet at night is a much greater inconvenience and expense than finding good blankets for your horses.

One of the most frequently asked questions in the early days of outdoor horse keeping was whether horses were cold living out in the winter, and the answer, more often than not, was: they're horses, not people... they're fine! But many caretakers, especially with purebred horses, saw the exact opposite: simply put, their horses were cold. It doesn't always require the coldest temperatures, but sometimes just a quick drop from warm to cool. Cold, wet weather can really have an impact on a horse's well-being. Wild animals have far more options when it comes to finding shelter from the wind and cold. They can find shrubs, tree cover, or uneven terrain to protect them from wind gusts.

Horses lose the most energy on wet, windy days. Even when temperatures are above freezing, rain and wind can cause horses to lose enough heat that they're unable to regulate their body temperature. Horses who can't maintain an adequate core temperature can suffer from serious health issues.

Maintaining an optimal body temperature is an absolute necessity for any warm-blooded organism. A constant body temperature ensures the normal functioning of bodily processes. The heat that a horse receives from his environment, together with the heat he produces, must be in equilibrium with output. If this equilibrium is disturbed, overheating or excessive cooling can make a horse very uncomfortable and even quite sick.

A horse's muscle mass is excellent at retaining the heat produced by processes like digestion (fermentation of fiber in the intestines) and producing heat through exercise.

A horse's core isn't subject to temperature changes based on his surroundings (because horses are warm-blooded), but the peripheral parts of the body (ears, hooves, skin) behave much like a cold-blooded animal's would: they partially adjust to the ambient temperature, helping to maintain a stable core temperature. This is why you can't tell whether a horse is cold by feeling his ears!

Temperature Management

Horses are excellent at regulating blood flow in the subcutaneous vascular network, which significantly influences temperature management of the entire body. Heat loss can be managed by vasoconstriction: blood flow is reduced in areas where blood would be cooled by being close to the surface of the skin.

Horses are also able to cool off through vasodilation: by expanding these same blood vessels, surface contact is maximized and excess heat is transferred through the skin.

If you aren't sure whether your horse is cold, place a hand by the elbow of a front leg. If this area is cold, your horse is cold enough that he's uncomfortable.

Skin, subcutaneous tissue, and fat act as powerful insulators. For this reason, it's a good idea to have horses go into the winter season with a few fat reserves—they will be far less bothered by the cold than a skinny or underweight horse.

Also, a well-made shelter can help protect a horse from the worst weather—an animal protected by a shelter can save up to 25 percent more energy reserves than an unprotected one. This is no small thing on difficult days! Horses that aren't as cold on a windy, heat-robbing day will consume hay at a more relaxed pace as well, saving some strain on your wallet. A well-thought-out shelter, whether you build new or rebuild by converting an existing building, is an investment which will pay you back over time.

Don't think that just because you're cold, your horse is cold, too! The human body loses heat much faster than a muscular horse. Plus, we as humans are spoiled—we're acclimated to seek out more and more warmth, while our horses are far better adjusted to the cold. If we lived outside all year, we would probably find a heated home unbearable.

Shivering

Some take a shivering horse as the definitive signal that it's time to blanket. Others believe that because shivering produces heat, the horses are warming themselves and are actually fine.

So What's the Real Story?
Peripheral thermoreceptors are activated by the skin becoming cold. As soon as a horse's usual defenses fail to maintain a comfortable temperature—hair bristling (increasing the coats' ability to insulate against the cold) and vasoconstriction (narrowing of the subcutaneous blood vessels)—the horse will begin to shiver. Using muscle tremors, the body will be able to increase heat production by about 30 percent. But there's a catch—this only works for a short time, and it uses a great deal of energy. Unfortunately, the first horses to begin shivering are usually somehow compromised: skinny, sick, old, or recovering from illness or injury.

These horses are the ones with the smallest energy reserves, but even a fat, healthy horse can't produce heat for long by shivering—the energy necessary for such demanding work is exhausted early. Although the body has other defenses against the cold as well (accelerated metabolism, for example), their effectiveness is limited. Shivering should be considered a sign that your horse is headed toward hypothermia—don't take it lightly.

Organs that produce the most heat:

- the liver, kidneys, heart, and skeletal muscle

The coldest organ:

- the skin

Despite horses being extremely well-adapted to the cold, if they're in a location where they have no shelter from the wind or rain, they will start to show signs of excessive heat loss during extreme weather. As we've learned, horses can generate heat very efficiently through muscle exertion and digesting fiber.

But even with free-choice hay available all day, a horse without any kind of windbreak can be virtually

Horse hair is such a great insulator during dry cold weather that it allows horses to happily roll around and also rest comfortably in the snow for several minutes. Of course, if an older horse lies down in the snow and can't get up, he'll get quite cold! Even wet fur can help keep a horse comfortable, thanks to the sebaceous glands. These glands produce sebum, an oil which serves to "waterproof" a horse's coat. Only if a horse becomes wet to his skin does his coat cease to fulfill its insulating function.

With constant movement, a large herd, and the ability to go to the basic enclosure to seek out shelter whenever they need to, these horses live very comfortably even on very cold days.

The basic paddock should serve not only as a mud-free space to save the pastures, but also as a reliable place to take shelter during extreme weather.

frozen by the weather. How should a horse warm up in this situation? By working? When a horse feels this cold, he'll stand in place, lost in himself, with his head down; he won't consider playing or even moving in this state. A group of horses may huddle together for warmth, but many horses are alone, or one of a pair; these horses can't benefit from the thermal help of many nearby bodies. Even a large group of huddled horses can't always warm themselves enough. So what should we do? Blanketing can be a solution, but we'll be running back and forth taking them off and putting them on. The best and most convenient solution is providing a functional, well-built basic shelter (which I may have already mentioned once or twice).

A good basic enclosure should have plenty of shelter options and places to escape the wind. Likewise, a pasture with natural windbreaks and dense vegetation that can also protect against rain can allow horses to make do with only their natural coat during the winter. If horses have these conditions met, they can be comfortable even when the temperature dips quite low.

In some areas, horses can manage temperatures as cold as -40°. Of course, a horse would need to be acclimated gradually to this kind of extreme cold, but horses who are accustomed to living outdoors tend to adjust much faster.

Nevertheless, there can always be a member of the herd that starts to show signs of being very cold: shaking, standing with his head down, basically telling you he's cold and uncomfortable. Even if he seemed fine just the other day in the same weather, it's senseless to do anything but help this horse get comfortable. Help him warm up by bringing him into a windless shelter and seeing to it that he isn't evicted by the other horses. If this won't work, find him a stall or other shelter.

Winter Riding—Handling a Winter Coat

A horse who lives outdoors will have a much thicker coat than one that spends his nights in a stall. I love that when winter comes, my lovely, refined Arabs are replaced by shaggy beasts. I can sleep well at night, knowing they're comfortable under piles of hair.

There's one huge catch, however: my horses are perfectly warm and comfortable standing around all day, but exercise under saddle can make them

These horses were brought into their basic paddock on an evening when, after a beautiful autumn day, the weather suddenly changed, bringing rain, snow, and wind. Even so, the sight of dinner coming brought the horses out from under the roofs. One of them seemed cold, so he was blanketed for the night. Giving him a bit of extra warmth for a few hours won't ruin his ability to self-regulate his temperature, as some may fear.

extremely sweaty. We have the advantage of dressing for the task at hand and can plan to wear less clothing if we're going to exercise. Horses with winter coats, however, can't switch to a light jacket. Imagine going for a run in a fur coat, a thick hat, and a heavy scarf—sound uncomfortable?

Some horses kept at show stables are clipped and blanketed. Hobby horses are worked only to the extent that they won't return home sweaty. Some are given the winter off entirely.

After a workout, I follow the principle of a thorough walking cool-down. I like to take young horses ponied next to riding horses. This way, the horses who exercised have a chance to dry off during easy movement, and the young ones get a little education. I never feel like I'm up against the clock to get a horse dried off, however, since I'm careful to finish before they're dripping with sweat.

You won't see horses like this living outdoors in the winter.

★ ★ ★

After a winter ride, never return a sweaty horse to the herd without the necessary care—there will be consequences for his health! If you don't have time after a 60-minute ride to spend another 30 minutes swapping wet coolers for dry ones, change the duration and amount of work so that your horse is completely dry or as close to it as possible when you're finished.

Everyone's routine will be a bit different, depending on their own priorities, preferences, and time constraints. Of course, the first priority is to be sure that our horses can face any weather conditions comfortably and in good health. It goes without saying that a clipped horse won't fare well.

They'll need increased supervision and above-standard care, including multiple blanket changes according to the weather. This is another reason horses often go without blankets if left outdoors—blanket changes with every temperature shift are unrealistic for most. If horses are blanketed, they're probably still unclipped and just need a bit of extra help, or they may have a (very) partial clip.

Most sport horses who work at the higher levels of their discipline won't be kept outdoors. One notable exception: endurance horses, who I often see being kept in outdoor situations even if they compete at the higher levels.

Studies show that a horse's thermoneutral zone—the temperature range where they don't have to expend energy to maintain a normal body temperature—is influenced by the prevailing temperatures of their environment and the duration of those temperatures. Data is often different: some literature reports a range from 0°–80°F, while others are considerably more conservative. Louda et. al., 2000 states: in long-term exposure to neutral and low temperatures, an animal's thermoneutral zone will expand and shift towards lower ambient temperatures.

For me personally, I've managed my horses to promote a thick, healthy winter coat. I prefer to vary the horse's work in the winter to avoid needing to blanket. How you manage your horse's exercise is up to you—but always pay attention to your horses. With the furry ones, take care to keep them comfortable under saddle, and for (clipped or) blanketed horses, be sure they're comfortable in their paddock.

There are several reasons to blanket. Maybe it's a one-time event to keep a horse warm during extreme weather, a daily necessity for a clipped horse, or you simply want to keep your horse clean to make tomorrow's grooming easier. For a horse in heavy work year-round, if he's blanketed, he'll have less hair and will be less likely to suffer health consequences from excessive sweating if he's worked indoors in the winter.

Reasons not to blanket:

- Your basic paddock is built so well, it essentially performs all the functions of a stall.
- A healthy horse going without a blanket will grow a heavy coat, ensuring he already has the best quality blanket available.
- Fewer blankets will need to be replaced and repaired—saving that money for something else.
- No need to obsess over weather reports or run out for constant blanket changes.
- Horses become more adaptable to fluctuations in weather.

Reasons to blanket:

- For one-time help on extreme days.
- When a horse has to be clipped.
- For performance horses. If they're blanketed consistently starting in the fall, they'll grow less hair, which will allow them to work through the winter without sweating excessively.
- To save time grooming.

In Conclusion: Should Every Horse Be Kept Outdoors?

I began this book describing the function of the basic enclosure and extolling the virtues of the basic paddock. That's how I'll finish as well. The very essence of a happy and comfortable relationship between horse and human is set up so that horses have enough freedom, and their humans can comfortably manage them. Thanks to a basic enclosure, this becomes easy to achieve. Regardless of the size of your space, first and foremost your goal should be to build a solid foundation for happy horses and people.

Very often, I hear from people who tell me they would love to put their horses out year-round but whenever they try it, their horses fail to thrive, leading them to believe that the only option for their horse is indoor stabling. I agree that not every outdoor living situation—especially a very basic one—is suitable for every horse. But I don't believe that a horse exists that wouldn't benefit from being able to go outdoors.

From a Stall Straight to a Year-Round Pasture?

A "lifer," thrown outdoors after a lifetime of living in a stall, can be a huge problem for owners and caretakers. A stalled horse learns to view the walls of his stall as security and protection, especially at night. If we push him straight out into a vast pasture with no shelters, we can inadvertently cause him all kinds of stress.

A horse who has been stalled for a long time won't be used to clouds of insects, all-day rain, unmuted sounds from all directions, and so on. Be considerate of the needs of these animals. He shouldn't be expected to make this transition immediately. His introduction to a new herd in particular should be done gradually and with care.

If your basic enclosure has a shelter that can be used to create a private space for a new horse, so he can see and touch other horses without being unable to get away from them (add some nice bonuses like treats to the space), he'll soon learn that the basic paddock is a safe space. The longer he uses it, the more he'll fixate on it, much as he once did when waiting to be brought into his stall from turnout at the end of the day.

If you're convinced you want to have your horses outside year-round, know that they can adjust to this new way of life—with a little effort. When you accommodate the initial needs of each particular horse, you'll be rewarded with a happy herd.

Who would disagree that horses look forward to going into their stalls in the evening? It's obvious that they're waiting by the gates as soon as the time approaches. What I see is an animal that has learned a habit, and habits can be useful to us as we train if we use them properly. Horses will look forward to going into any space where they get extra benefits, like a favorite treat or being cooled off on a hot day. And what would make them happier than, if when they're done licking their dinner bowl clean, or they've finished having a nice nap in the straw, they could leave whenever they chose? One of a horse's favorite pastimes is walking here, there, and everywhere.

Horses always return to where they're most comfortable. If we encourage habits, we can create an extra level of security for our horses. Fixation on a common living space is shaping behavior—we can encourage this behavior by providing all the necessary benefits (the company of their herd mates, a safe space, clean water and hay, feeders, shade, shelter from wind and rain...) to ensure the basic paddock becomes more valuable to a horse than a stall could ever be.

Remember that every living creature is happy to return to their favorite place, and for a horse, the happiest place is to be safe and comfortable with his herd mates.

Good luck and get building!
Iveta Jebáčková-Lažanská

Custom Housing

Customizing group housing for horses sounds difficult, especially in commercial stables. And it's genuinely not easy to accommodate individuality within a collective. The following few rules will help you:

- Build shelters with an eye toward easy division into boxes, or divide the interior space. This way, the newcomer in the herd can get away for a bit and give himself some space.
- To start with, place new individuals in a smaller herd with friendly, problem-free horses.
- If possible, place horses of approximately the same weight and body type in the same herd, so you can adjust the amount of hay and straw to suit all of them at once.
- Be prepared to readily give extra doses of trough feed to a new horse if his condition requires it; set aside a fenced area inside the basic paddock for this purpose.
- Don't be afraid to limit outdoor horses from time to time; it won't hurt them at all to get used to more variation in their lives. It is sad when a horse has to change owners and suffers at first in his new home, simply because he never once had to adjust to a small paddock and box in his entire previous life.

Photo Credits

Many thanks to all who were willing to share their photos for this book!

Chapter 1:
opening photo: Lenka Brothánková
p. 1: Iveta Jebáčková-Lažanská
p. 2: Lenka Brothánková
p. 3: Eva Ftorková
p. 4: Helena Zahradníková
p. 5: Aktivstall Grasbrunn
p. 6: Dana Barochová
p. 7: Carina Terhorst

Chapter 2:
opening photo: Lenka Brothánková
p. 10: Renata Pulchartová
p. 11: Kerstin Namenich
p. 12: Elisabeth Ernst
p. 13: Inge Ludwig
p. 15: Aneta Svobodová
p. 16: Veronika Sixtová
p. 17: Iveta Jebáčková-Lažanská

Chapter 3:
opening photo: Slunná stáj Říčany
p. 20: Eliška Trpíková
p. 21: Maja Kupčáková
p. 23: Dana Barochová
p. 24: Markéta Nacházelová, Iveta Karasová, Aneta Novotná
p. 25: Marie Vosiková
p. 26: Michaela Svobodová, Veronika Myslivečková
p. 27: Hof Etzenbach
p. 28: Podještědí Farm
p. 29: Lenka Brothánková
p. 30: Jaroslav Míč
p. 31: Sinah Worm
p. 33: Iveta Jebáčková-Lažanská
p. 36: Hof Vosswinkel
p. 37: Sinah Worm
p. 41: Jana Steidl-Kindernayová

Chapter 4:
opening photo: Lenka Brothánková
p. 44: Lenka Brothánková
p. 45: Klára Bišická
p. 46: Eva Miškechová
p. 48: Aktivstall Steinloge
p. 49: Gawsworth Track Livery
p. 50: Renata Marková, Iveta Jebáčková-Lažanská
p. 52: Veronika Zámečníková, Gawsworth Track Livery
p. 53: Lenka Brothánková

Chapter 5:
opening photo: stock.adobe.com
p. 55: Veronika Huda
p. 56: Michaela Šeflová
p. 57: Hof Vosswinkel
p. 58: Anna Havel, Kateřina Mrázová
p. 60: Martina Cerhová
p. 61: Lenka Brothánková
p. 62: Aneta Novotná
p. 63: Jiřina Jandová
p. 64: Lenka Brothánková
p. 66: Lenka Brothánková, Aneta Novotná
p. 67: Renata Štiebrová
p. 68: Eva Baarová
p. 69: Michaela Šitnerová
p. 70: Klára Bišická, Meike Stricker
p. 71: Zuzka Medková
p. 72: Veronika Albrechtová
p. 73: Aneta Novotná

Chapter 6:
opening photo: Aneta Novotná
p. 76: Lenka Brothánková
p. 77: Lenka Brothánková, Grindstone Paddock Trail
p. 78: Hof Etzenbach
p. 79: Michaela Šitnerová
p. 81: Cornelia Klarholz

Chapter 7:
opening photo: Lenka Brothánková
p. 84: Zykloopenhof
p. 85: Stáj Vlčice, Michala Hersteková
p. 86: Milena Nováková, Manuela Brandtner, Hana Jelínková
p. 87: Statek Rezkovi, Šárka Arnoštová, Aneta Novotná
p. 88: Carina Terhorst, Lenka Brothánková
p. 89: Horse Options, Eva Machovcová, Feelution Paddock Trail
p. 90: Stáj Dalbett
p. 91: Gawsworth Track Livery
p. 93: Jasmin Lange, the author's archives, Aneta Novotná
p. 94: Hof Vosswinkel Paddock-Trail, Annette Rosso
p. 95: Mary-Theres Haller

Chapter 8:
opening photo: Aneta Novotná
p. 98: Sarah Markwalder-Vogler
p. 99: Gawsworth Track Livery
p. 100: Statek Rezkovi
p. 101: Vanessa Leitner, Gudrun Gläser-Bleyer
p. 102: Anubi Triedre, Christine Schneider
p. 103: Kingsland Ranch, Natalie Zoe

Chapter 9:
opening photo: Iveta Jebáčková-Lažanská
p. 106: Jana Novotná
p. 107: Equus Bohemia, FreizeitstallAsenbeckPaddock Trail
p. 109: Gut Heinrichshof
p. 110: Iveta Jebáčková-Lažanská
p. 112: Lenka Brothánková
p. 113: Dörte Kolkmeyer

Chapter 10:
opening photo: Aneta Novotná
p. 116: stock.adobe.com
p. 117: Lenka Brothánková, Aneta Novotná
p. 118: Lenka Brothánková
p. 119: Iveta Jebáčková-Lažanská
p. 120: Iveta Jebáčková-Lažanská, Jitka Rauerová
p. 121: Lenka Brothánková

Chapter 11:
opening photo: stock.adobe.com
p. 124: Iveta Jebáčková-Lažanská
p. 125: Eva Ftorková
p. 127: Iveta Jebáčková-Lažanská
p. 129: Zykloopenhof
p. 130: Iveta Jebáčková-Lažanská
p. 131: Eliška Křivánková
p. 132: Lenka Brothánková
p. 134: Aneta Novotná
p. 136: Iveta Jebáčková-Lažanská
p. 137: Lenka Brothánková
p. 138: A Little Paradise Ranch
p. 139: Lenka Brothánková
p. 140: Aneta Novotná
p. 142: Lenka Brothánková

p. 143: Niki Rypová
p. 144: Aneta Novotná
p. 146: Lenka Brothánková
p. 148: Iveta Jebáčková-Lažanská
p. 149: Lenka Brothánková
p. 151: Iveta Jebáčková-Lažanská
p. 152: Aktivstall Steinloge
pp. 153, 155–156: Lenka Brothánková
p. 157: Anne Meuschke
p. 158: Iveta Jebáčková-Lažanská
pp. 160–161: Aneta Novotná
p. 162: Julia Hein
p. 163: stock.adobe.com

Chapter 12:
opening photo: Iveta Jebáčková-Lažanská
p. 166: Iveta Jebáčková-Lažanská
p. 167: Lucie Kovaříková

Chapter 13:
opening photo: stock.adobe.com
p. 170 illustrations: Lída Slavíková
p. 171: Dana Barochová
p. 172: Lucie Perlová
p. 174: Eva Smrčková
p. 175: Martina Součková

Chapter 14:
opening photo: stock.adobe.com
p. 178: Lenka Brothánková
p. 179: Yvona Jelínková Formanová
p. 180: Gawsworth Track Livery
p. 181: Eva Chadimová
p. 183: Renata Štieberová
p. 185: Silke Juppenlatz

Chapter 15:
opening photo: Maayke Langedijk
p. 187: stock.adobe.com
pp. 188–189: Kathjana Möller-Wendell
p. 190: Jana Nováková
p. 191: the Czech company Přírodní ploty [Natural Fences], Eva Smrčková
p. 193: Maayke Langedijk

Chapter 16:
opening photo: Aneta Novotná
p. 196: Iveta Jebáčková-Lažanská
p. 197: Natural Paddock (Spain)
p. 198: Zdena Skopalová
p. 200: Iveta Jebáčková-Lažanská
p. 201: Lenka Brothánková
p. 203: Iveta Jebáčková-Lažanská

Chapter 17:
opening photo: Lenka Brothánková
p. 206: Mary-Theres Haller
p. 207: Iveta Jebáčková-Lažanská

Chapter 18:
opening photo: Renata Štieberová
pp. 210–213: Aneta Novotná
pp. 214–215: Manege zonder Drempels
p. 216: Sigrid Layer, Laufstall-hollergarten; De Paardenmaat (Holland)
p. 218: Amelia Phillips, Janina Kröger
p. 219: Lenka Brothánková
p. 220: Adriana Kozárová, Lenka Brothánková
p. 221: Horse Naturally boarding for horses
p. 222: Mary-Theres Haller
p. 223: Iveta Jebáčková-Lažanská, the Caesar Neslovice estate

Chapter 19:
opening photo: Aneta Novotná
p. 225: Andra Jemand
p. 226: Lenka Brothánková, Tereza Marešová, Michaela Šitnerová
p. 227: Iveta Jebáčková-Lažanská, Lena Boettge, Grindstone Paddock Trail, Christa Dennstedt

Chapter 20:
opening photo: Ranč Vránov, Marcela Kozová
p. 229: Adéla Volfová, Marcela Kozová, Aneta Novotná
p. 230: Likit Liz from Waldhausen, Aneta Novotná
p. 231: Petra Hrubá, Aneta Novotná, Zykloopenhof

Chapter 21:
opening photo: Dana Barochová
p. 234: Horse Naturally boarding for horses
p. 235: Tereza Marešová
p. 236: Eva Miškechová
p. 237: Paddock-Trail Hof Freimann
p. 239: Glandy Connemara
p. 241: Grindstone Paddock Trail
p. 242: Gawsworth Track Livery
p. 243: Mary-Theres Haller
p. 244: Christiane Trost
p. 245: Tereza Marešová

Chapter 22
opening photo: Aneta Novotná
p. 248: Equus Bohemia
p. 249: Simone Wik
p. 250: Natural Paddock (Spain)
pp. 252–254: Iveta Jebáčková-Lažanská
p. 255: Zykloopenhof
pp. 256–257: Iveta Jebáčková-Lažanská

Chapter 23:
opening photo: stock.adobe.com
p. 260: from the author's archives
p. 262: Małgorzata Mąkosa
p. 263: Iveta Jebáčková-Lažanská
p. 265 photo: Sigrid Layer; illustration: Lída Slavíková
p. 266: Iveta Ruszova Červenková
p. 267: Jana Novotná, Lucie Hronová, Veronika Sixtová

Chapter 24:
opening photo: Lenka Brothánková
p. 270–271: Iveta Jebáčková-Lažanská

Chapter 25:
opening photo: Marie Bulvová
p. 274: Renata Štieberová
p. 276: Jana Sotonová, Jana Steidl-Kindernayová
p. 277: Glandy Connemara
p. 278: Lenka Motyčková

Index

Page numbers in *italics* indicate illustrations.

Y

Z